An Ahrensburgian Site at Zonhoven-Molenheide (Belgium)

Pierre M. Vermeersch

BAR International Series 2471
2013

Published in 2016 by
BAR Publishing, Oxford

BAR International Series 2471

An Ahrensburgian Site at Zonhoven-Molenheide (Belgium)

ISBN 978 1 4073 1082 4

BAR Publishing is the trading name of British Archaeological Reports (Oxford) Ltd.
British Archaeological Reports was first incorporated in 1974 to publish the BAR
Series, International and British. In 1992 Hadrian Books Ltd became part of the BAR
group. This volume was originally published by Archaeopress in conjunction with
British Archaeological Reports (Oxford) Ltd / Hadrian Books Ltd, the Series principal
publisher, in 2013. This present volume is published by BAR Publishing, 2016.

Printed in England

BAR
PUBLISHING

BAR titles are available from:

          BAR Publishing
          122 Banbury Rd, Oxford, OX2 7BP, UK
EMAIL   info@barpublishing.com
PHONE   +44 (0)1865 310431
  FAX   +44 (0)1865 316916
          www.barpublishing.com

IN MEMORY OF CHRIS PELEMAN, WHO
INVESTED SO MUCH OF HER ENERGY AND
HER "SAVOIR FAIRE" IN THE EXCAVATIONS
AT MOLENHEIDE AND IN THE REFITTING
OF THE MOLENHEIDE FLINTS

CONTENTS

FIGURES

TABLES

INTRODUCTION

The site of Zonhoven-Molenheide was discovered by Mr Roger Maes who had on a regular basis been carrying out a survey of an area that was then being used as a training field by the Belgian army. As part of their training the soldiers had dug pits for the purpose of defence. It was in the redeposited sand from these pits that Roger Maes first noticed archaeological artefacts. He organised a small excavation and collected a total of 435 artefacts by sieving the sediments from the pits and their surroundings.

Following contact with the then Laboratory for Prehistory at the *Katholieke Universiteit Leuven*, Belgium, it was decided that the site should be protected from further destruction by organising a programme of field research. Therefore the Laboratory for Prehistory, in association with the *Provinciaal Gallo-Romeins Museum* at Tongeren and the tourism administration of the municipality of Zonhoven, organised a series of excavation campaigns with the aim of exploring the whole site. This project was led by Pierre M. Vermeersch.

A total of nine excavation campaigns have been organised.
06/07/1993 tot 30/07/1993 with Didier Verbruggen as field supervisor.
11/07/1994 tot 05/08/1994 with Chris Peleman as field supervisor.
12/07/1996 tot 02/08/1996 with Chris Peleman as field supervisor. 03/07/1997 tot 25/07/1997 with Chris Peleman as field supervisor.
13/07/1998 tot 06/08/1998 with Chris Peleman as field supervisor.

12/07/1999 tot 29/07/1999 with Chris Peleman as field supervisor.
17/07/2000 tot 28/07/2000 with Chris Peleman as field supervisor.
16/07/2001 tot 28/07/2001 with Chris Peleman as field supervisor.
15/07/2002 tot 27/07/2002 with Chris Peleman as field supervisor.

These excavations were carried out by both students and volunteers. The project could also count on the benevolence of the municipal and military authorities responsible for the area.

We would especially like to thank Roger Maes, who first introduced us to the site, Guido Pirotte, who as cultural representative of the Zonhoven municipality was always ready to solve practical problems, and Guido Creemers, who as representative of the *Provinciaal Gallo-Romeins Museum* at Tongeren provided constant support .

We would also like to thank Veerle Rots for writing chapter 6 and Peter Tomkins for correcting the English.

We would like to thank Elena Marinova for her observations and for her inspection of the charcoal and Etienne Paulissen, who corrected chapter 1. All errors remains the author's.

Artefact drawings were produced by Gunther Noens (for the nice ones) and by the author.

1 - GEOLOGY AND GEOMORPHOLOGY

1.1 – Geomorphology

The Zonhoven-Molenheide site is located in the wooded area of Molenheide near Zonhoven (5.42253°N 50.99460°E; Kadaster Sectie D 3e blad J 24n23), Limburg, Belgium (fig. 1, site 1). It is situated on a rather flat area at the western edge of the Kempen Plateau, south of the Roosterbeek river. The site is situated at about 76 m asl and 25 m above the valley bottom of the Roosterbeek river.

The Kempen (Campine) Plateau is situated in the Belgian province of Limburg. It is a fan-shaped plateau with a gentle slope to the north. Its altitude decreases from about 100 m in its southernmost part (Lanaken) to about 30 m near the Dutch border. Its southern and eastern border is characterised by a well-marked and steep slope along the Meuse Valley. A relatively abrupt transition can be observed from the Plateau to the low lands in the west. The descent of the western plateau edge is formed by slope erosion (Frederickx, Gouwy 1996).

The western edge of the Kempen (Campine) Plateau in intensely dissected by river valleys, which developed perpendicular to the plateau edge. These valleys are remarkably wide and in many places marshy. West of the Kempen Plateau the Glacis of Diepenbeek-Beringen is a NW-SE oriented band that gently descends from 50 m in the NE to 35 m in the SW. The gentle slope of the erosion glacis links the flood plain of the Demer in the South with the steeper slope of the Kempen Plateau. The surface of this area is slightly undulating as a result of incisions by rivers, such as the Roosterbeek and the Slangbeek which drain the plateau. The Roosterbeek, which has a wide valley bottom, lies just to the north of the site.

1.2 – Geology

The Kempen Plateau is topped by fluvial deposits, which are the remains from terrace deposits of probable Middle Pleistocene age, that were created by the Meuse and the Rhine rivers. These deposits rest on sand of the mid-Miocene Bolderberg Formation (fig. 2), which consists of heterogeneous white and brown grey quartz sand of continental and marine origin. The deposits on the Kempen Plateau are considered to be the deposits of a braided river, fundamentally different from the deposits of the present Meuse. On the basis of their characteristics, it is assumed that these fluvial deposits are periglacial in origin and probably of Middle Pleistocene age. At the Molenheide site this terrace deposit belongs to the "Zanden van Winterslag" (Paulissen 1983, Gullentops *et al.* 2001). These gravely sands are Meuse deposits which include a large amount of reworked local material of the Bolderberg Formation mainly at their base. On the Kempen Plateau, late Pleistocene cover sands are very thin and even absent in several places. Their base is disrupted by cryoturbations (Frederickx, Gouwy 1996).

West of the site, on the slope towards the Roosterbeek, there is a surface where the local water table seeps out of the plateau flank creating a restricted area of peaty marsh (fig. 6). However, coring with an augur in this marsh suggests that the peaty deposit is of recent age as it is underlain by a dry podzol.

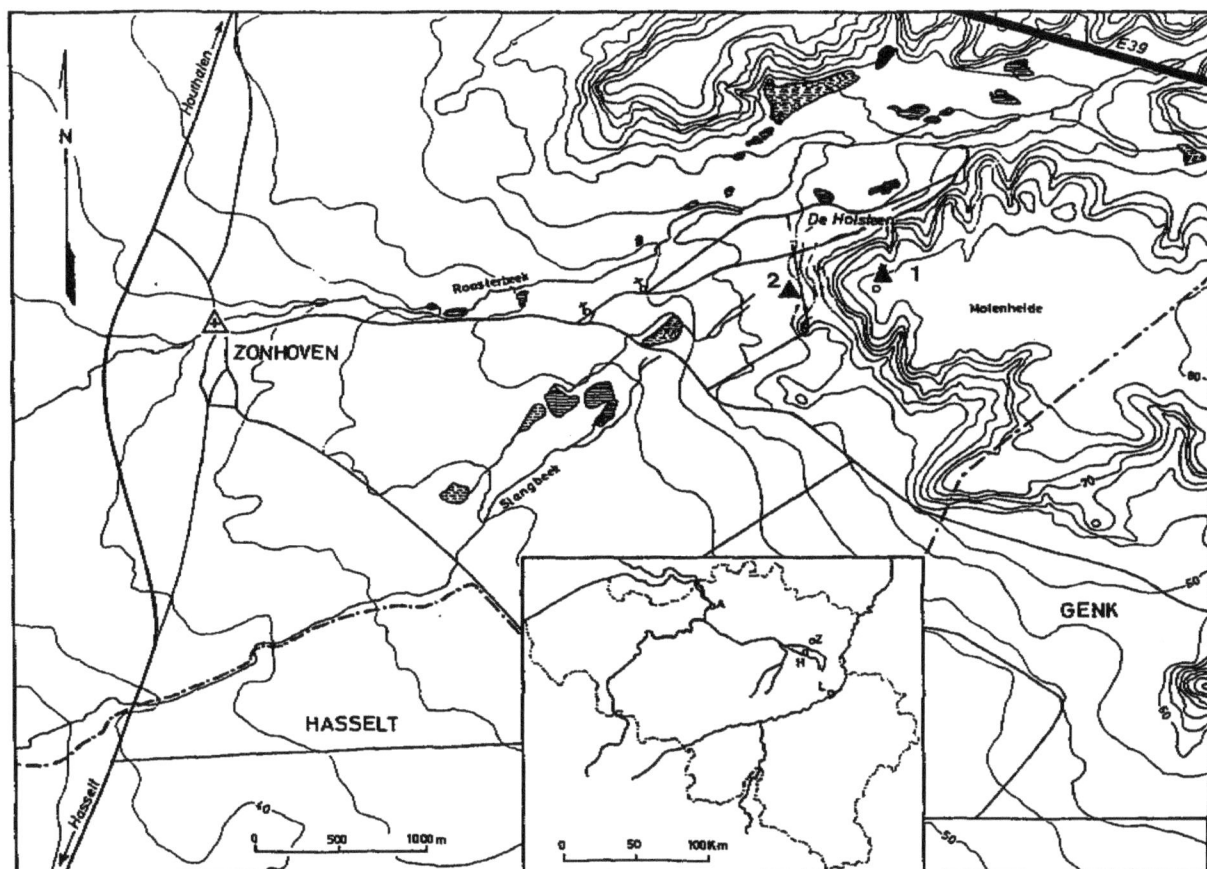

Figure 1 - A map of the area of the sites: 1: Zonhoven-Molenheide; 2: Zonhoven Kapelberg.

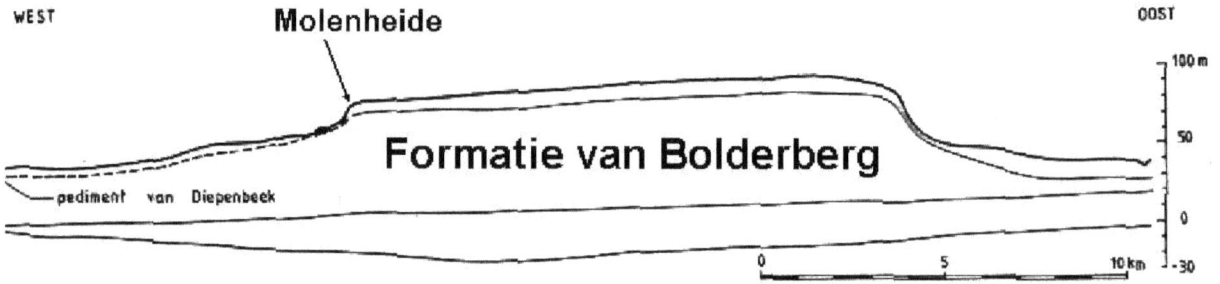

Figure 2 - General geological section of the area showing Molenheide on the edge of the Kempen plateau.

The surface cover in the area of the site, which is marked as a heather area on the 18th century Ferraris map, is still more or less preserved thanks to its incorporation into a training camp of the Belgian Army. It appears never to have been exploited agriculturally, but has suffered from intensive pit digging by soldiers who utilised the area for military manoeuvres. The landscape has now a vegetation cover of secondary open wood with birch and pine (fig. 5, 8).

1.3 - Excavation

The aim of the project was to gain an understanding of an Ahrensburgian site and to determine the exact limits and mutual relations between the different exposed artefact concentrations. Originally, it was thought that there were several sites present within the area (Vermeersch, Creemers 1994). For that reason the earliest recovered artefact concentrations were named Zonhoven Molenheide 1 and Zonhoven Molenheide 2 (Peleman, Vermeersch, Luypaert 1994). Later (Vermeersch *et al.* 1996; Vermeersch, Peleman, Maes 1998; Peleman, Vermeersch 2002) this internal site numbering system was abandoned when it

appeared likely that the two concentrations belong, not to two isolated sites, but to a single large artefact distribution, in which several discrete artefact concentrations occur.

A topographical survey of the site and its environment was performed, resulting in a topographical plan of the site itself (with contour intervals at 0.1 m) and of and of the slope down to the Roosterbeek (with contour intervals at 1 m) (fig. 6). The absolute elevation of our reference point was estimated by comparing it to the elevation indicated on the 1:20,000 topographical map of Zonhoven. Our local datum of 10 m is presumed to be about 76 m a.s.l. This means that the absolute elevation of our local datum is only an estimation.

Grid construction commenced from two perpendicular lines, magnetic south-north and west-east. The intersection of these two lines was considered as the 0-point of the grid (fig. 6). Each point or surface within the research area was identified on the basis of its distance from this 0-point along the two cardinal axes. Two iron rods, situated at 00N00E (10.66 m relative elevation), and 50N00E (8.63 m relative elevation) were set up permanently at the site.

Figure 3 - Google earth image of the site showing the position of the topographical plan (see figure 6).

12

Figure 4 - Digital Terrain Model of the Molenheide area (courtesy of the VIOE).

Figure 5 - General view on the site during the excavation.

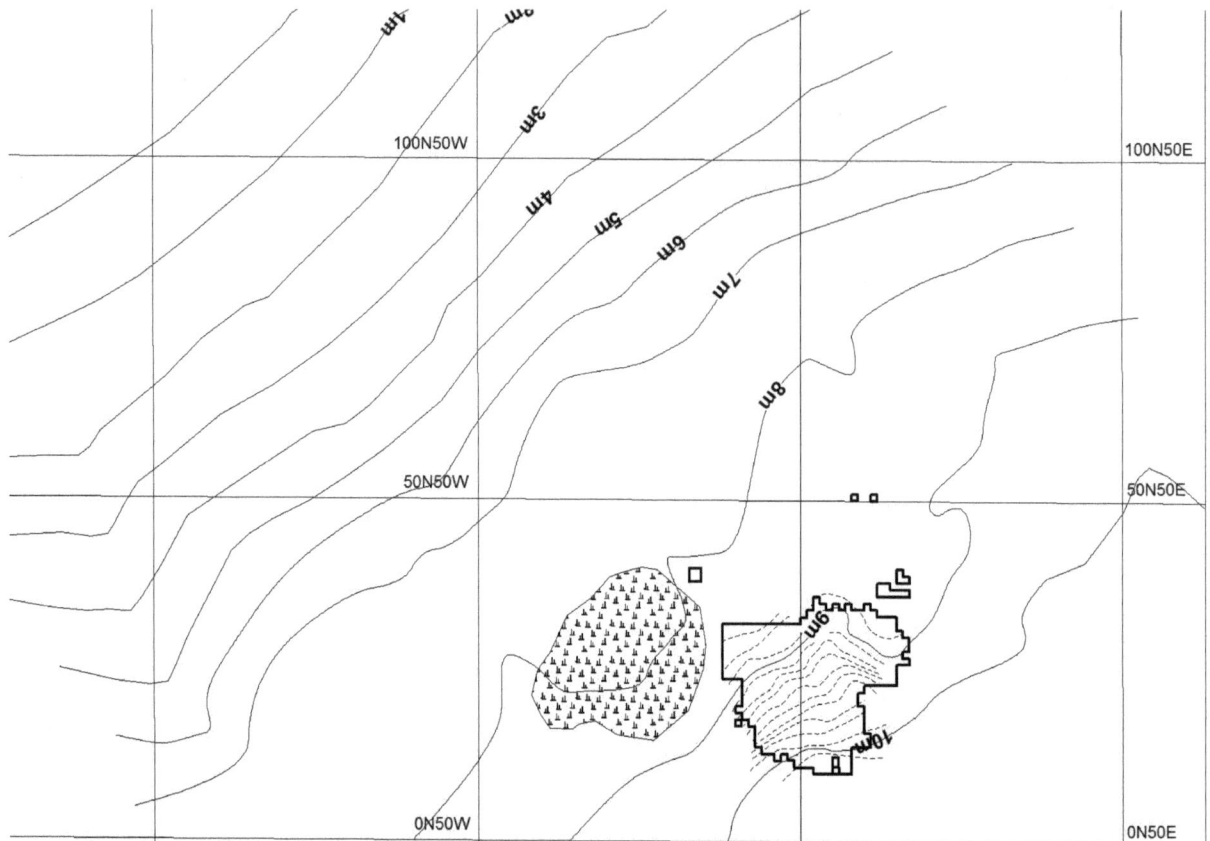

Figure 6 - Plan (square on fig. 3) of the excavated area and the peaty swamp on the slope down to the Roosterbeek.

The excavation was organised in squares of 1 m2. Since it took 9 excavation campaigns to cover the entire area, over the years the excavated units became somewhat scattered over the site's surface (fig. 7).

For the uncovering of artefacts a trowel was used in areas with a significant find density and a shovel in areas with a low find density. Initially squares of 2 x 2 m., spread across the area, were deployed. Later it proved to easier to use trenches of 1 x 1 m. During the first excavation campaign all artefacts were measured in three-dimensionally. The excavation dump was sieved using a 5 mm screen. However, this proved to be very time consuming, especially because not all excavators were sufficiently aware of the distinction between a real artefact and a natural frosted flint pebble fragment, the latter being very numerous in the sediments.

From the second excavation campaign onwards, a different method was applied, where each square metre was divided into four quarters, each one quarter of a square meter. Each quarter square meter was excavated, mostly using a shovel to remove spits, each approximately 5 cm thick. Each spit was sieved and the elevation of the recovered spit was recorded. In this way we could control the vertical position of the artefacts while registering a sufficiently precise horizontal artefact distribution (Cziesla 1988).

Artefacts were numbered in the field using the site initials (ZM) followed by the year of the excavation (e.g. 93) and the square coordinate (e.g. N1205), and then a slash followed by an individual identification number (i.e. 1 and following). Each artefact was registered on an excavation form. In later years the "O" for "Oost" in Dutch was replaced by E (East). In later excavation years the square coordinate identifier was replaced by a simple number. If that number is not followed by a slash, the artefact was

recovered from the sieve without a precise location in the square or quarter square.

The artefact distribution plan, used in this report, is the result of a mixture of precisely registered artefacts and artefacts registered only within their quarter square metre or square meter. In the distribution plans less precisely registered artefact positions were artificially distributed over the relevant surface using randomisation software (Microsoft Excel©) to generate coordinates. In the plans no distinction has been made between locations provided by specific measurement and those that are the result of randomisation.

The eastern part of the excavation comprised an area that was entirely destroyed by a large pit (fig. 17). Here we did not register the exact place of each artefact. Nevertheless an important artefact concentration was found in the fill of the western part of this pit. In the refitting plans for this concentration the artefacts were randomly scattered in the square where they were found. The fill of the pit still had large lumps of the upper soil horizons preserved, which suggests that here, unusually digging did not seek to break up these upper soil horizons. Initially we suggested that this destruction was caused by the activity of the army tanks that used the area. However the layout of the pit and the preservation of the soil lumps can better be explained by the activities of "loam diggers", which are known to have been active in the area. "Loam diggers" were searching for the fine alluvial deposits present in the coarse sands and gravels of the Kempen plateau for the purpose of house construction[1]. For that reason they dug pits, broadly the size of a room, through the gravel layer to reach the loamy deposits beneath. The activity of such

---

1   Information provided in 1995 by Dr. M. Dusar of the Belgian Geological Service

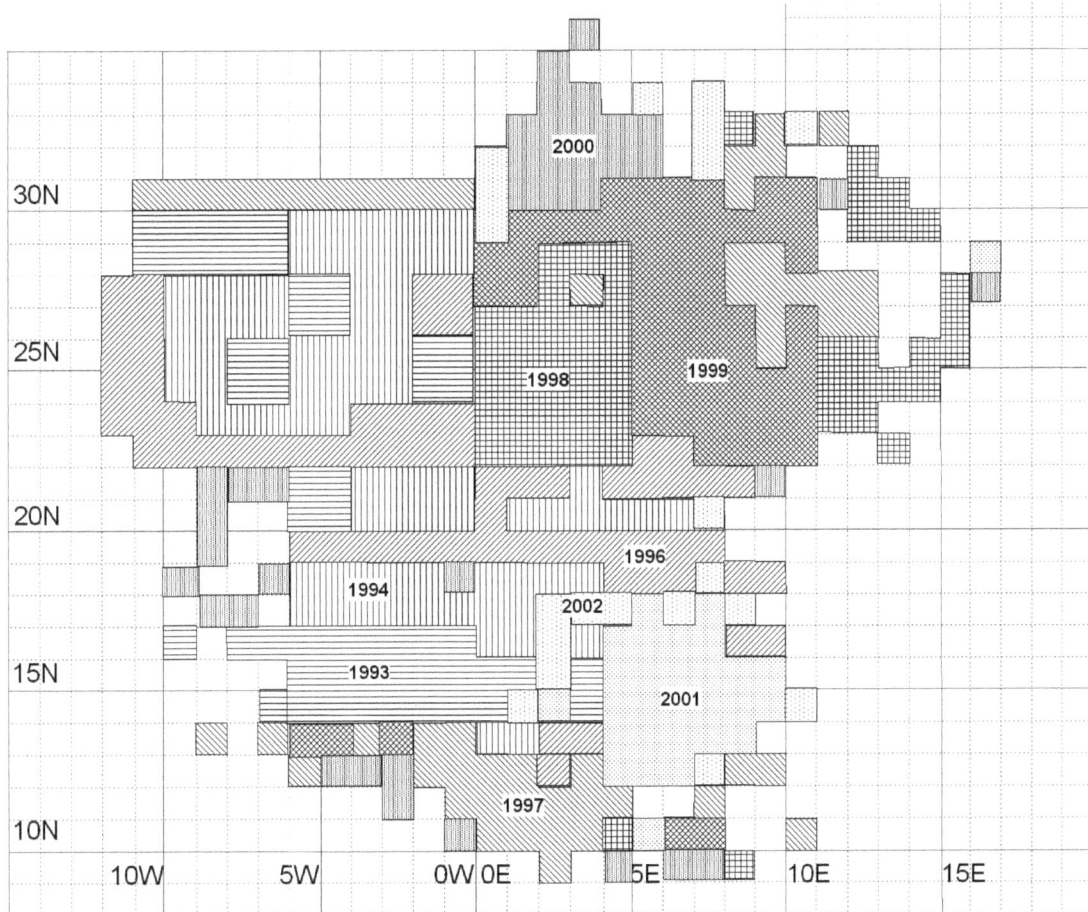

Figure 7 - Squares excavated during each year of excavation.

Figure 8 - View of the site at the start of the excavations; Chris Peleman (†) is holding the surveyor's rod.

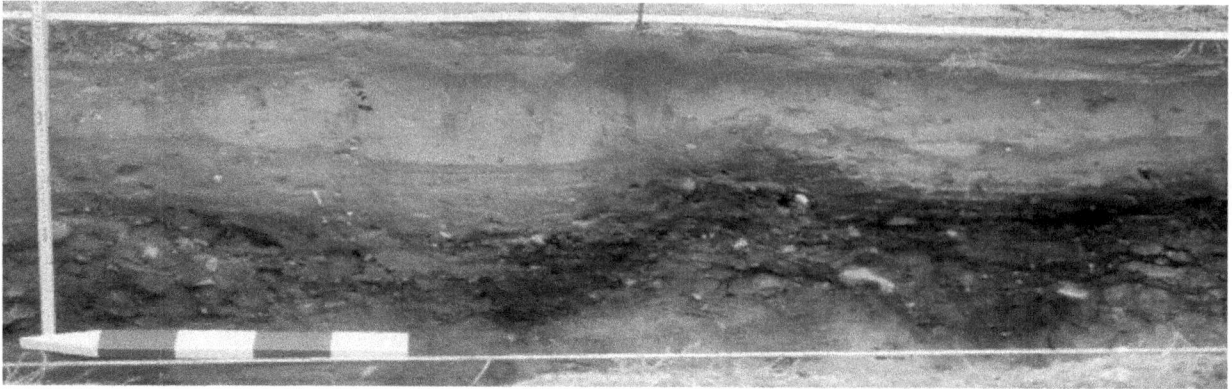

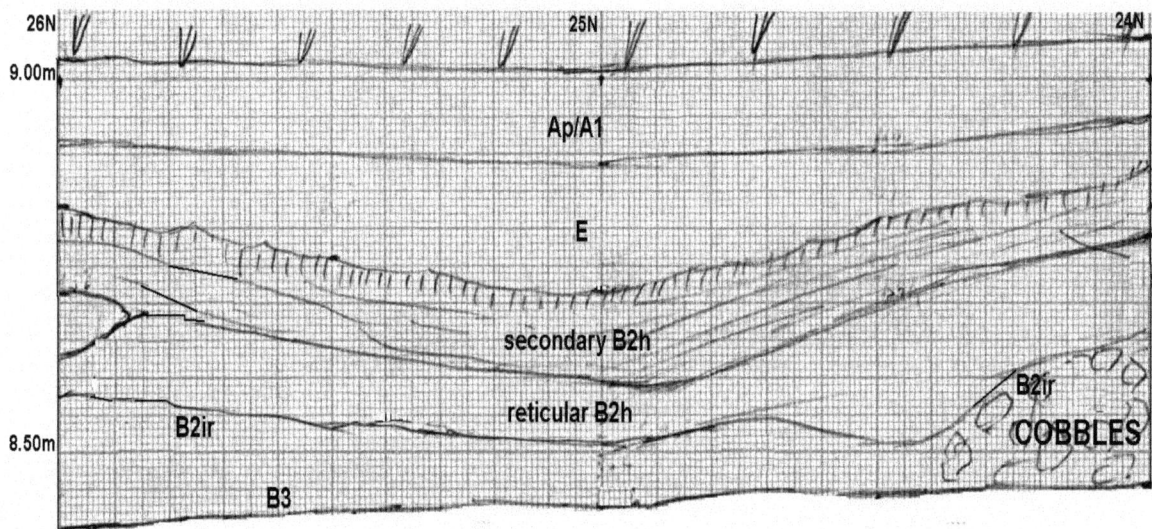

Figure 9 - Section 24-26N05W, the left part of which is shown also as a drawn section.

loam diggers is probably also reflected in the presence of wheel tracks in the excavated area (see below).

During the years of excavation additional data were recorded, which resulted in a rather patchy distribution map of the structures (artefact and gravel concentrations). We are confident that even this limited and patchy presentation will allow the reader to understand the distribution of the cobble-gravel layer (fig. 16-17).

1.4 - Stratigraphy

1.4.1 - Profile description

All over the surface of the site a humic iron podzol (orthic podzol) (fig. 9) had developed. There were no traces of ploughing which would have destroyed the upper part of the soil. Indeed, the area appears never to have been agriculturally exploited and consequently the original A-horizon is usually quite well preserved. Currently a humic-iron podzol is present, in which very few traces of faunal activity can be detected. During the excavations no traces of soil burrowing animals have been encountered and the soil seems to be devoid of animal life. Roots are concentrated in the A1 and in the Bh-horizons.

Figure 10 presents the lay out of a profile at 26-27N05W. The matrix of this profile is medium to fine sand with scattered fine river gravel. In this matrix a podzol was formed with a well developed A0 and an A1 of loose structure with an important raw humic accumulation, composed of grains

with a humic coating but also those that are white washed grains. In this profile and more widely the A0 and A1 (2.5 YR 5/1) are preserved. Only occasionally they have been reworked into a thin Ap horizon. The E-horizon is often rather thick, up to 30 cm, and mostly somewhat pink (5 YR 7/1). Sometimes, in its lower part, a darker secondary humic illuviation of about 5 cm thick had developed. Its upper limit is sharp. The B2h is mostly quite dark (5 YR 5/1 – 5YR 2/1) and only slightly consolidated. It is reticular and it has a sharp delimitation with the underlying hard pan, which is a 10 cm thick dark brown iron placic horizon (5 YR 3.5/3), in which some roots and rootlets are present. The placic horizon often coincides with a significant gravel and cobble accumulation. In some cases however the cobble and gravel accumulation is absent or occurs at a lower level. The B2ir below is often reticular and dark brown at its top becoming yellowish (10 YR 6.5/4) in lower levels characterised by numerous thin illuviation hairs. In most profiles small gravel of variable granulometry is present in the soil horizons.

Below the uppermost soil horizons, traces of an older brown soil, maybe a leptic podzol, are still visible characterised by numerous traces of animal burrowing activity.

The west profile at 26-27N05W (fig. 10) was sampled for granulometric analysis. This analysis (fig. 11) suggests that the sand fraction of the upper five horizons presents a quite similar particle size profile (medium to fine sand) with a mode of 2.26 φ. The upper five horizons are clearly homogenised, probably due to an significant bioturbation

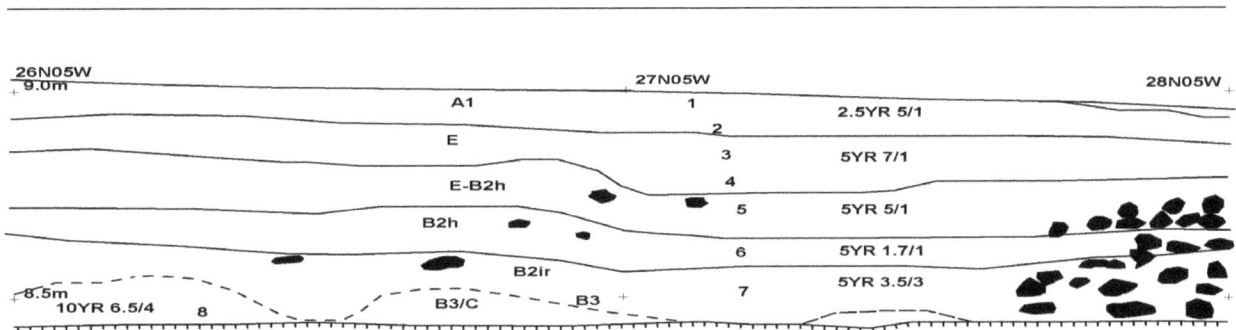

Figure 10 - The west section at 26-27N05W.

during the soil formation processes. Samples 6 (B2h) and 7 (B2ir) are somewhat coarser corresponding to the presence in the profile of more pebbles and blocs. In sample 8 (B3/C) the fine sand fraction is more significant.

Some excavation squares were dug to a lower level in order to improve our understanding of the site stratigraphy. At 21N05-06E, (fig. 12) the stratigraphy is as follows from top down:

1: Black raw humic fine to medium sand, which in the western part of the profile had been removed. This is the folic horizon (0) rich in decomposed plant remains of a podzol.

2. A slightly pink albic horizon in fine to medium unstructured sand. This is the eluvial (E) horizon of a podzol, with occasional small gravel (<2 cm) and artefacts.

3. Grey black humic fine to medium sand. This is the il-luvial spodic horizon (B2h) of a podzol, with occasional small gravel (<2 cm) and artefacts.

4. Black slightly consolidated fine to medium sand. This is a hard pan placic horizon (B2h) of a podzol, with numerous root traces, occasional small gravel (<2 cm) and artefacts.

5. Brown fine to medium sand with numerous sometimes very large cobbles. Together with horizon 6 it forms the B2ir of a podzol. The B2ir presents at its top a continuous brown colour which in lower levels degrades into brown sometimes hair thin fibres.

6. Beige brown fine to medium sand with some rare cobbles.

7. Convoluted fine to medium sand, very rich in small cobbles forming a vertical wedge in the lower horizons 8 and 9.

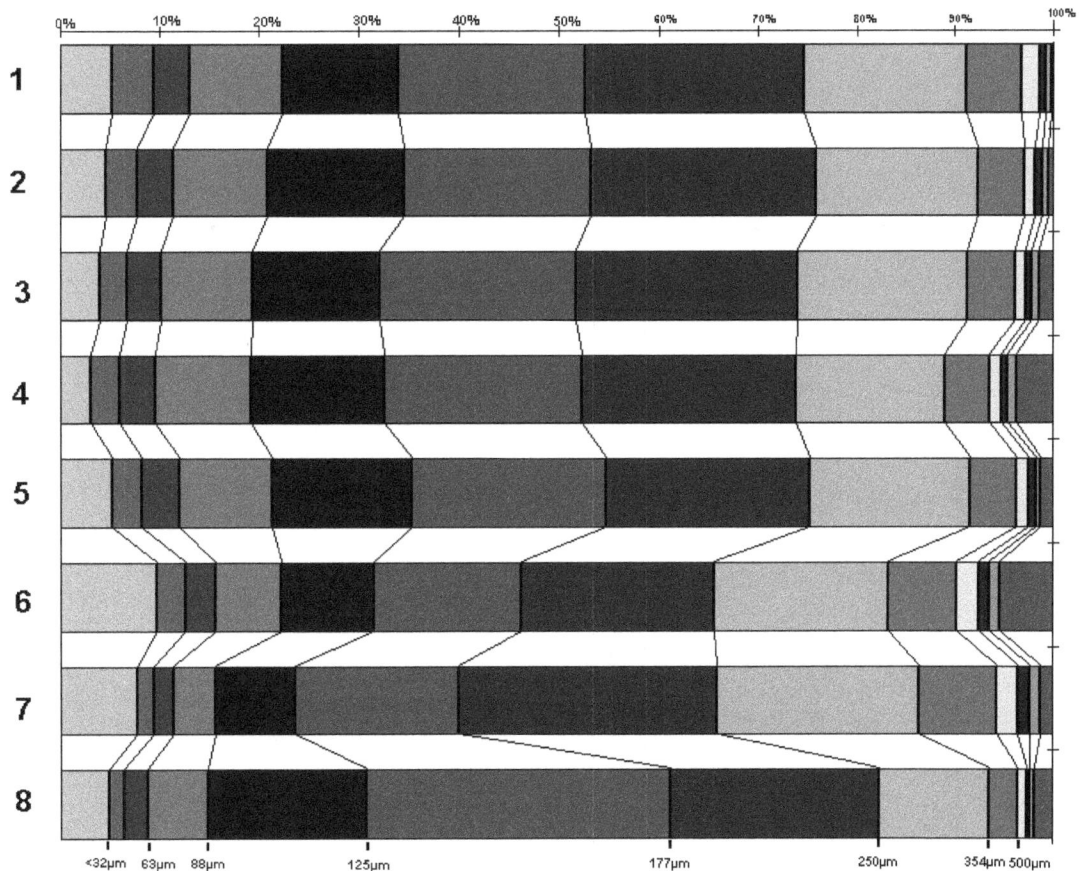

Figure 11 - Granulometry by weight of the fine fraction (< 2mm).

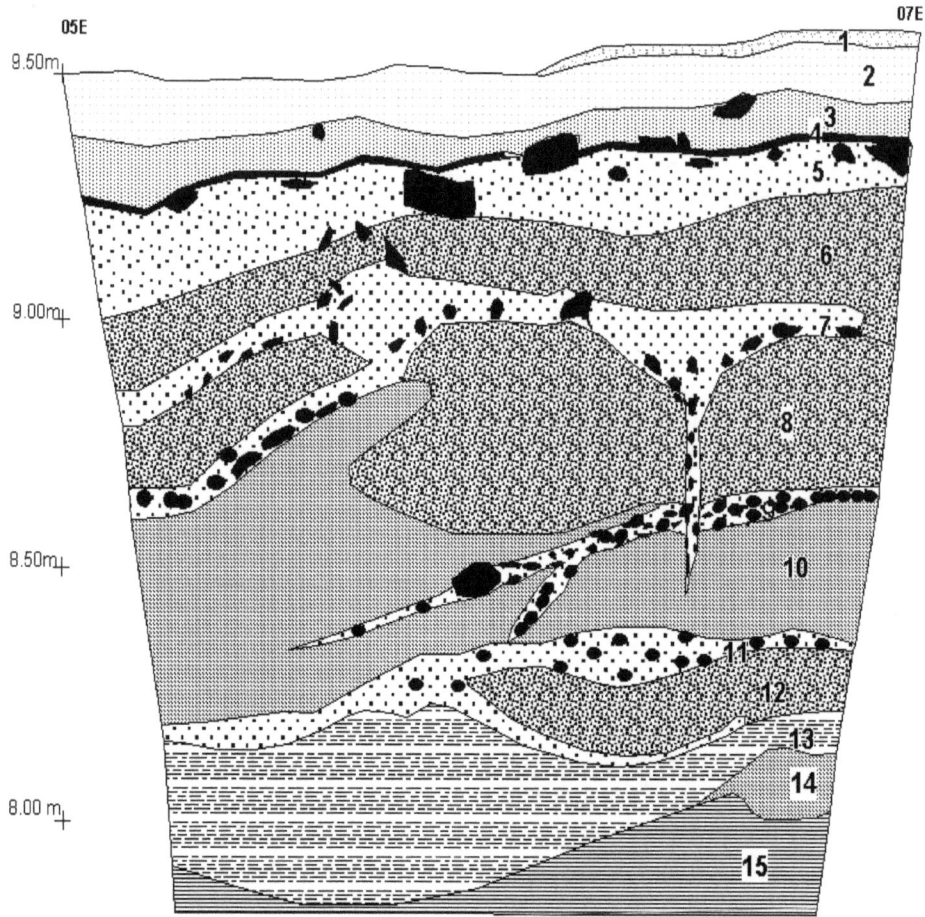

Figure 12 - Section at 21N05-06E (for legend see text).

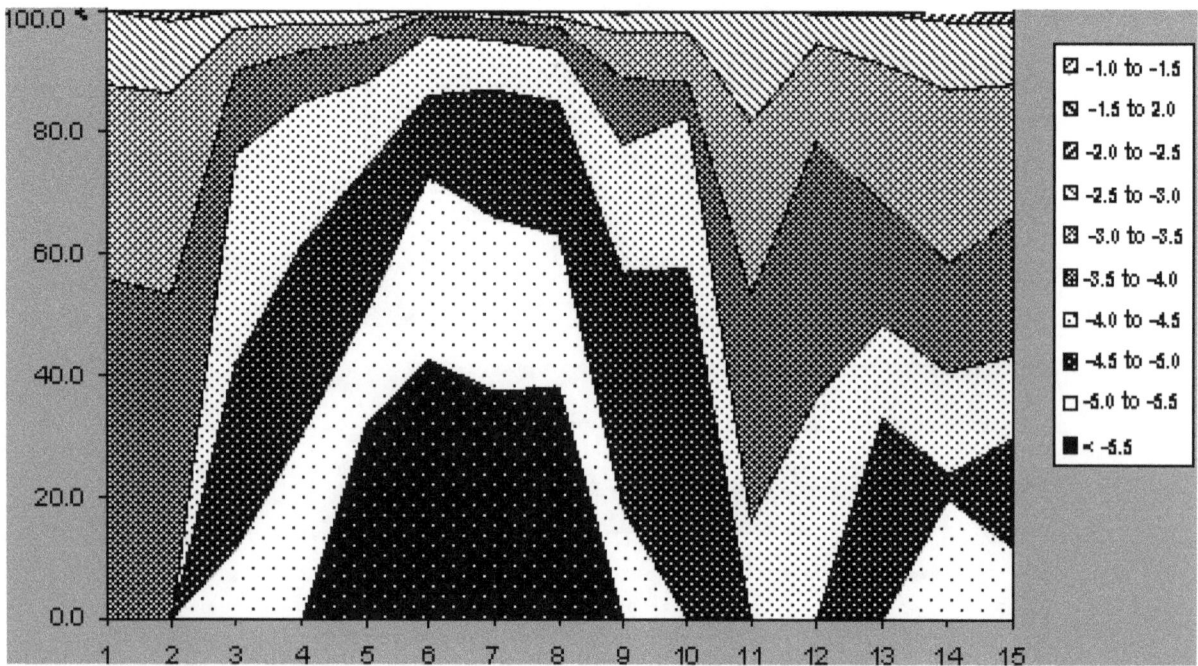

Figure 13 - Grain size analysis by weight (in φ) of fractions larger than 2 mm by 15 spits of 5 cm from top (1) to 75 cm (15).

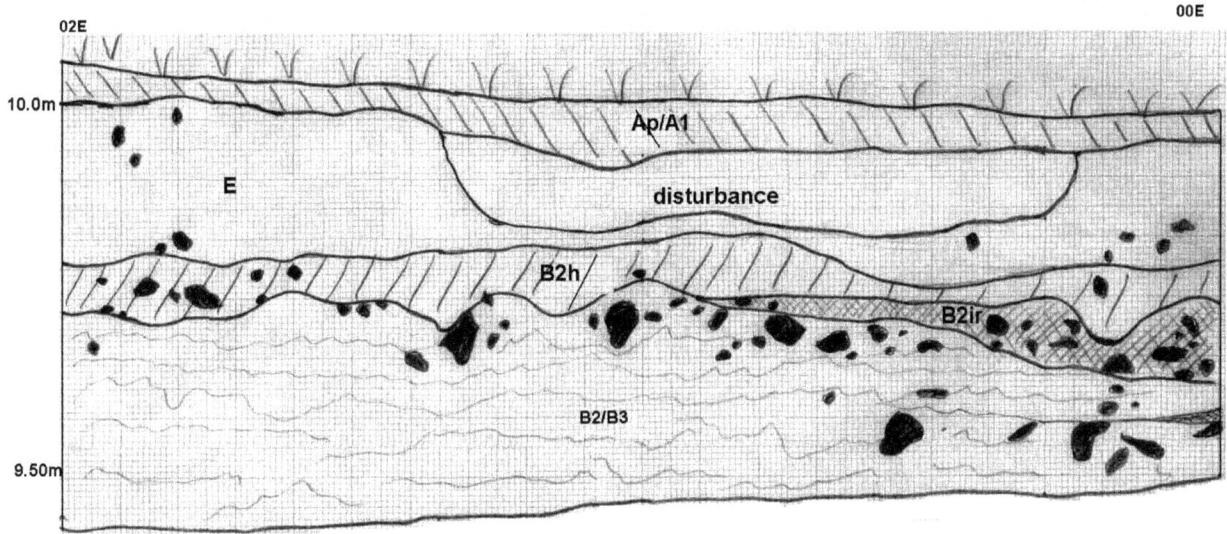

Figure 14 - South section of 13N00-02E with the presence of cobbles indicated.

8. Partially gleyified and partially whitish fine to medium loamy homogenised sand.

9. Convoluted gravel with occasionally some large cobbles in a fine to medium sandy matrix.

10. Slightly layered gleyified loamy fine to medium sand with thin clayey strata, becoming very rich in clay at its base.

11. Convoluted whitish fine to medium sand with gravel.

12. Orange layered fine to medium sand.

13. Stratified clayey fine to medium sand and sandy day.

14. Loamy sand.
15 Sandy clay.

The granules and pebble fraction was analysed by dry sieving and the results (fig. 12) confirm observations in the field evidence that the upper 10 cm contains only pebbles of the fraction -2.5 to -4.5 φ. In contrast the levels, from about -10 (3) down to -50 cm below the surface contain a significant amount of larger pebbles. Below -50 cm the pebbles are smaller. Most pebbles are rolled quartz, apparently derived from the gravel that belongs to the main Middle Pleistocene terrace deposits of the Kempen Plateau.

The profile is typical for the whole excavation area, although the thickness of the upper fine to medium sand var-

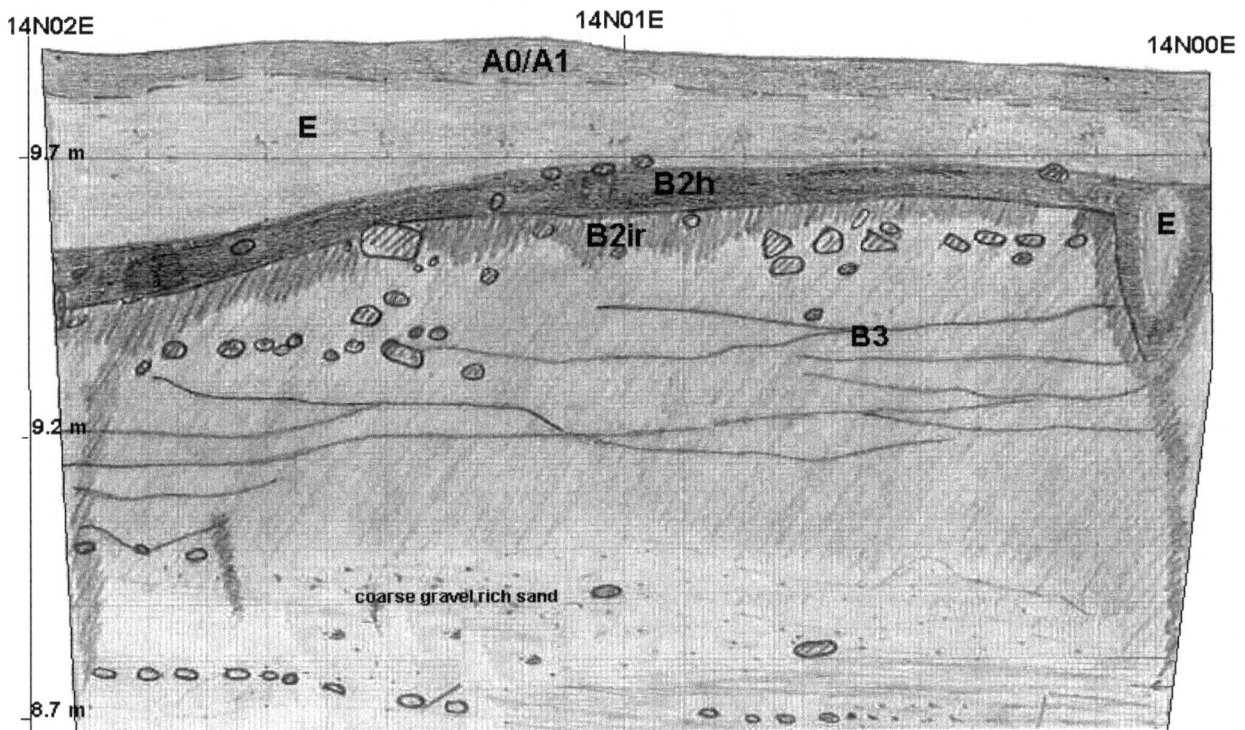

Figure 15 - Section 14N00-01E, where the gravel layer is clearly situated below the B2h-horizon.

Figure 16 - Cobble layer: sometimes dense (upper row); other times more scattered (middle row); detail of large aeolised blocs, some of which have been positioned upright (lower row).

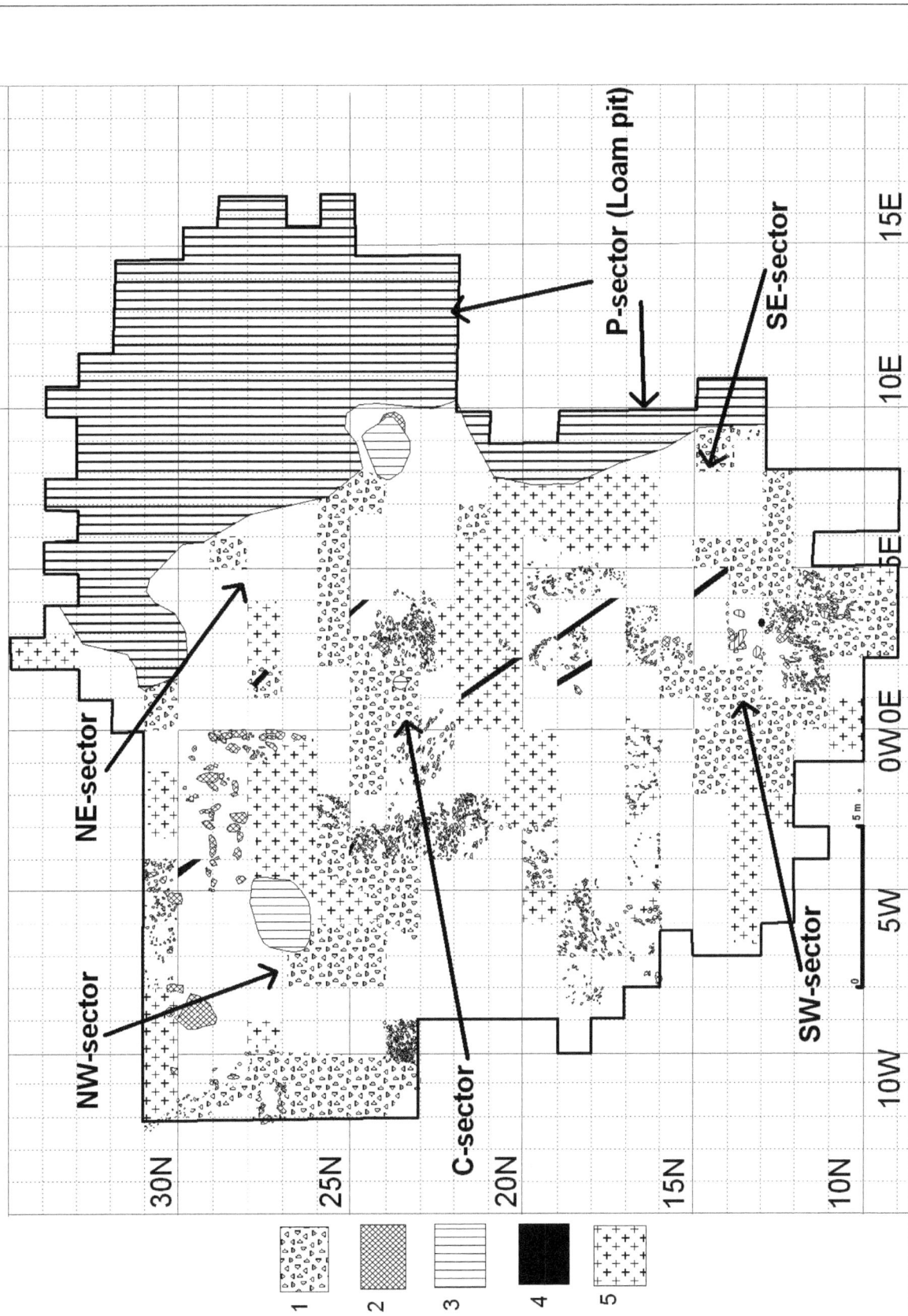

Figure 17 – Distribution plan of the cobble-gravel layer: 1: cobble layer present near the surface; 2: individual cobbles belonging to the cobble layer; 3: disturbance; 4: cart track; 5: absence of gravel layer

ies from 20 cm to about 100 cm and always rests upon fluviatile cryoturbated deposits (fig. 12).

In the south profile of 13N00-02E (fig. 14) the thickness of the E horizon is quite typical. A disturbance of the E horizon may be observed while the B2ir horizon is restricted and replaced in the east by a reticular B2/ B3. The accumulation of cobbles starts at about 40 cm below the surface.

However in some places, mainly in the southern and the north eastern part of the excavated area, the E-horizon has been truncated or is even absent, probably because of recent human activity. The B2ir top most often coincides with the top of the cobble accumulation, if the latter is present.

1.4.2 - Cobble and gravel distribution

The excavated area is characterised by the presence of a sometimes dense cobble and gravel layer (fig. 16-17), which is not, however, continuous over the whole area. It is situated at the base of the upper fine to medium sand and tops the fluviatile deposits underneath. Most often its surface is rather regular and more or less parallel with the present surface.

There are some areas where no cobble-gravel layer is present. It was observed in some profiles that several cryoturbated cobble-gravel layers occur on lower levels (fig. 12,15). The intense aeolisation of the upper cobble layer suggests that it is a lag deposit resulting from an intense deflation of outcropping cryoturbated cobble-gravel layers.

During the earlier excavation campaigns little attention was given to the distribution of that cobble-gravel deposit. Later, the presence of the cobble-gravel layer was recorded in more detail, to the extent even that detailed drawing of the individual cobbles were made. Such a variable degrees of recording resulted in a plan of the presence and distribution of the cobble-gravel layer (fig. 17) where the resolution is quite different from place to place.

During a field visit Frans Gullentops[2] made the following observations on the large cobbles of this gravel-cobble layer: "Nearly all the larger cobbles (> 10cm), but also many smaller ones, are aeolised. We have drawn and examined some quadrants. It is striking that most of the aeolisations occur at the upper side of the cobbles, and with a constant wind direction from the North East sector. This means that these cobbles have not moved since. Other cobbles have a fresh aeolisation above (on top of an older one,) and remnants of an earlier at the under-side, suggesting that those cobbles have been moved since the first or during the aeolisation period. The aeolisation resulted in a strong polish but without producing facets, suggesting it may be surmised that aeolisation did not last long or was with fine silt particles. The wind direction is in accordance with the NE winds that brought in the younger cover sands and the Brabantian loess during the dry end of the LGM. It is not in accordance with the SW winds that blew up the parabolic dunes of the Younger Dryas.

Some other cobbles are upright; the aeolisation is present but has not affected their exposed upper part, suggesting that they have been put upright after the aeolisation process. From the irregular situation it may be concluded that the upright cobbles are not the result of a cryoturbation process, but maybe due to a biological process as later

hooves pressure. It is remarkable that the cobbles not always rest on the same layer (horizon), sometimes on a humic aeolian grit layer which seems to be of a relative young age."

Such an interpretation suggests that the sand covering the upper cobble-gravel layer was already in place before the Younger Dryas. Indeed, the Younger Dryas south-west wind did not affect the top of the cobble-gravel layer.

In the excavated area the gravel layer is interrupted by some more or less circular structures, such as at 11NO2E, 24-25N04-05W, 27N1 1W (fig. 9-10), at the edges of which the gravels slightly dip towards a gap in the centre. The gap in the gravel cover is probably the result of cryoturbational activity with resulted in the formation of a non-sorted circle. This is one of the patterned ground forms described by Washburn (1973), where the material of the central areas in some forms has risen from depth. Such non-sorted circles have average diameters of 0.5-3 m, which fits the diameter of the gap in the gravel cover at our site. The formation of the gap is related to conditions prevailing during the last Ice Age, anterior to the formation of the aeolisation traces. There is no evidence to suggest that the layout of the gravels is related to the posterior human occupation of the area.

1.4.3 - Vertical artefact distribution

An attempt was made to register and study the vertical distribution of artefact at the site of Zonhoven-molenheide.

During the excavation it became dear that where the cobble-gravel layer is present (fig 15), artefacts are scattered from the present surface (the original Al horizon of the soil) down to the top of the cobble-gravel layer (fig. 18, 20), which coincides mostly with the B2h-horizon of the humic iron podzol. When the cobble-gravel layer is absent or less compact (fig. 18), artefacts have a much more significant vertical distribution and can reach a depth of up to 120 cm below the top of the Al horizon.

There is no doubt that the vertical artefact distribution is related to the presence of the cobble-gravel layer. Refits from a restricted surface area (fig. 20), such as artefacts belonging to Refit 142 or Refit 181, exhibit a significant vertical distribution, suggesting that artefact depth is not related to a succession of deposits covering artefacts which were produced during a single event. Other refits present a similar distribution. This vertical distribution is apparently the result of postdepositional activity. Moreover the sand in which the artefacts are found is apparently older than the presumed age of the human occupation, which, as will be discussed later, is correlated with the Ahrensburgian during the Younger Dryas. It may, therefore be suggested that the occupational remains were originally deposited on the surface of the upper sand layer.

Postdepositional processes seems to be responsible for the present position of the artefacts in the sandy deposits (Vermeersch 1976, 1977; Barton 1987).

Wood and Johnson (1978) have stated that "a reasonably accurate assessment of the pedoturbatory history of the soils and sediments at every archaeological site is absolutely prerequisite to valid archaeological interpretation". That statement is entirely correct, yet in reports on sites, where an important vertical artefact scatter is observed, frequently, no attempt is made to understand what the original occupation horizon was during the Epipalaeolithic and the Mesolithic periods.

---

2 We thank Prof. Frans Gullentops for this field visit and the interpretation of his observations.

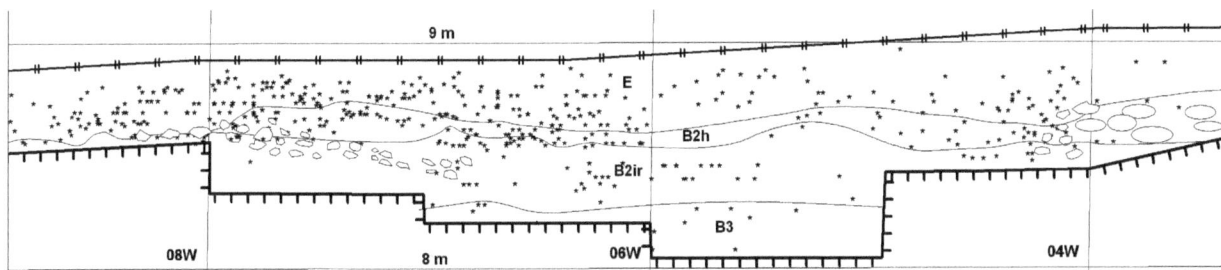

Figure 18 - Section 26N 3-9W through one of the gaps in the cobble-gravel layer, illustrating the important vertical scatter where no thick cobble-gravel layer is present.

It is remarkable that on the Zonhoven-Molenheide site the artefacts occur from the present surface, which still coincides with the original surface of the Al soil horizon at the end of the last Ice Age. In general artefacts, which exhibit a vertical distribution below the plough zone in sandy soils at Epipalaeolithic and Mesolithic sites in the sandy areas of Western Europe, are often interpreted as the result of bioturbation. At such sites, the original early Holocene surface was destroyed by erosion or by ploughing. The ploughing horizon prevented the observation that artefacts occur down from just below the early Holocene surface.

Balek (2002) insists that burial of all artefacts in stable soils, which were developed in pre-Holocene sediments, is due to vertical movement of the artefacts in response to normal biologic activity. It appears that, on such sites, the archaeological material is no longer in its original position and should therefore not be considered to be in a primary context. Rather it has undergone a movement down from the surface due to bioturbation and has also suffered a horizontal displacement of unknown magnitude.

It seems that on many sites in sandy Flanders no sediment accumulation has occurred since the onset of the Holocene.

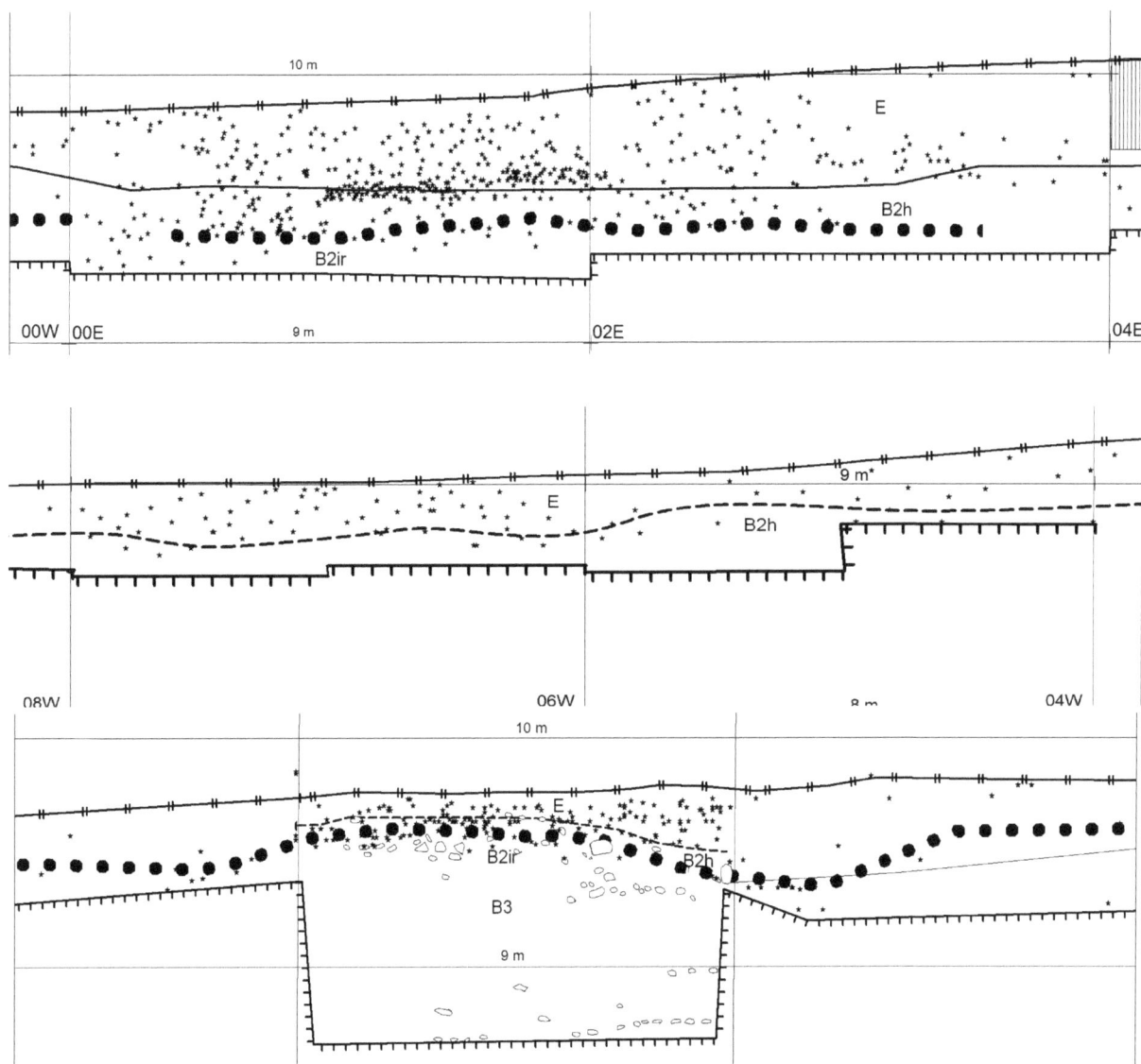

Figure 19 - Section at 12N00-04E (upper), 23N04-08W (middle) and 14N00W-03E (lower).

If that is correct, we have to accept that at such sites the Epipalaeolithic, the Mesolithic and all later occupation horizons coincide with the presentday soil surface. From that original surface the remains migrated down during a period when the present humic iron podzol had not yet developed. Very often, at such sites, including Zonhoven-Molenheide, remains of an ancient brown podzolic soil are observed. In such a soil, artefacts could continue to be moved down by bioturbation. In the BC-horizon of that soil, which is often found below the B2ir-horizon of the present soil, numerous animal burrows can be observed. Further downwards movement of artefacts was only prevented when a iron podzol and a humic/iron podzol developed . In our region, such podzolisation occurred mainly during the late Neolithic and the Bronze Age (Munaut 1967).

At Molenheide, the vertical artefact distribution down from the top of what was originally the surface of the landscape at the end of the Last Glacial suggests that artefacts, originally deposited on that surface, later moved down because of postdepositional processes.

If there was enough time between the occupation and the destruction of the original surface - a situation that fits most of the Epipalaeolithic and Mesolithic sites - then artefacts may already have been moving down into the soil horizons and may consequently have been (partially) been preserved from later destruction.

The question is, of course, when and how this bioturbation could have occurred as the soil, at least in its present state, is almost entirely devoid of burrowing animals. High magnification micro wear analysis has been performed on the artefacts, which, macroscopically, are in a very fresh state of preservation. Use wear was present but cannot always be easily recognised because of the presence of significant postdepositional weathering of the flint surface. This weathering is due to friction from sand particles on the artefact surface (Rots 1996, see also page 81). This surface weathering suggests that the artefacts underwent an significant movement within the deposits. We thus must accept that significant postdepositional vertical displacement of the archaeological material has taken place at Zonhoven-Molenheide (Vermeersch 1999, Vermeersch & Bubel 1997). Horizontal displacements, however, may well have been more restricted.

If there was re-occupation, as may often have been the case, at the same site during the Early Holocene, the remains from both occupations would have moved down, eventually collapsing into a single artefact horizon. This could result in the mixing of archaeological remains from successive occupations. We presume that this was indeed the case at the site of Brecht-Moordenaarsven 2, where remains of a Middle and a Late Mesolithic occupation are found were mixed (Vermeersch, Lauwers, Gendel 1992). A similar situation might lie behind the presence of some Wommersom quartzite artefacts on the eastern part of the site at Zonhoven-Molenheide.

At the site of Zonhoven-Molenheide, the presence of a cobble-gravel layer had an significant impact on the vertical distribution of artefacts. When the cobble-gravel layer is present, it prevents vertical artefact movement. When the gravel layer is absent, artefacts could move much further down.

In 15-16N06E artefacts occur at an significant depth below the surface. This situation seems to be the result of a tree-fail. This phenomenon was not observed during the excavation and owing to the excavation method in quarter square meters, the artefact distribution cannot now be checked against the distribution pattern typical for such events (Crombé 1993).

In 14N05-06E a tree fall is probably responsible for the irregularity of the gravel layer and the occurrence of artefacts at a much lower than in the surrounding squares.

1.4.4 - Interpretation of the stratigraphy

The lower deposits on the site consist of coarse fluviatile sands, sometimes homogeneous, sometimes mixed with patches of fluviatile gravels. They probably belong to Middle Pleistocene terrace deposits. The upper part of the sandy deposits shows traces of cryoturbation. The top of the fluviatile deposits mostly comprises a discontinuous layer of large cobbles. Above this an apparently thin layer of cover sands is present on which substrate a humic-iron podzol with a humic or humic/iron B-horizon ("Zcg" - and "Zbg gronden" on soil map) developed with its characteristic horizons. It is obvious that the eluvial horizon is more pinkish in colour than the eluvium from a podzol developed on aeolian sand under heather vegetation.

The sequence of events, from early to late, should be as follows:

• During the Middle Pleistocene deposition of the "Zanden van Winterslag" by the Meuse river.

• Inversion of the relief creating the Kempen Plateau, the western border of which is eroded by the river Nete and its tributaries

• During the later Pleistocene the Plateau surface iundergoes erosion and a lag deposit of gravel and cobbles of Meuse origin froms over the top.

• During periglacial climatic events the lag deposit is submitted to cryoturbational activity leading to the formation of non-sorted circles.

• During the Brabantian at the end of the late glacial maximum (26.5 - 14 ka BP), during which winds were predominantly nort-east, the surface of the cobble layer was aeolised and subsequently covered with a thin layer of cover sands.

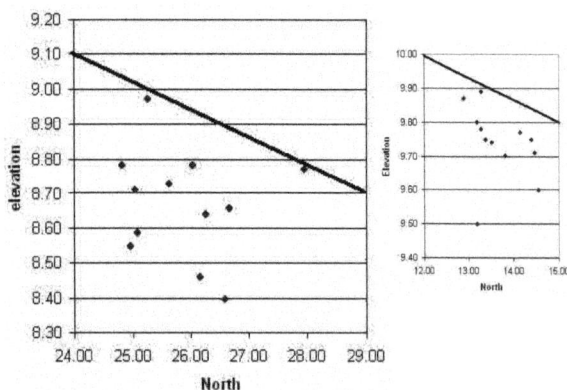

• During the late glacial period humans visit the site and leave their artefacts on the surface.

Figure 20 - Vertical position of artefacts from Refit 142 in the NW-sector (left) and Refit 181 (right) in the SE-sector in relation to the present surface (thick line).

• After humans visitation these artefacts remains rest on the surface and deflation certainly results in the removal of light material such as charcoal and chips.

• Mainly during the early and middle Holocene, under forest cover (Munaut 1967), a brown forest soil is formed. Bioturbation results in the vertical descend, and scattering of the human artefacts. Bioturbation is mainly restricted to the upper coversands because of the presence of the cobble-gravel layer and normally does not affect the lower fluviatile deposits. The remains of a brown soil attests to conditions that were less acid than those of the present. Such conditions existed all across the sandy area prior to human clearance of the original Atlantic forest, from the Bronze age onwards, and the subsequent formation of heath vegetation (Munaut 1967). This resulted in an in-

creased eluviation of raw humus and the formation of the present orthic podzol (Scheys, Dudal, Bayens 1954). In the BC-horizon of that pre-Bronze Age soil, which often occurs below the Bir-horizon of the present soil, numerous animal burrows can be observed.

After deforestation (in the Bronze age?) a heather is installed across the surface, resulting in an significant acidification of the soil.

A humic-iron podzol is formed under the heather. The formation of the B-horizon is to a large extent determined by the depth of the earlier bioturbation resulting in a cor-relation between the B-horizon and the cobble-gravel layer. The human artefacts are distributed in the A-B-horizons of the podzol.

2- STRUCTURES

2.1 - Distribution of the artefact concentrations

Several artefact concentrations may be defined and numerous refits between the different concentrations can be observed (fig. 21-22). No stratigraphic arguments are available to differentiate several multiple occupations in the excavated area. Refits between the different artefact concentrations suggest that all collected artefacts could belong to a single human occupation. For that reason in the present study the collected artefacts and their distribution are treated as resulting from a single occupation event.

Within the excavated area six distinct artefact concentrations are evident: one to the north west (NW-sector), a smaller one to the north east (NE-sector), one to the south west (SW-sector) and one nearby to the south east (SE-sector). In between these there is a fifth, less well-defined central concentration (C-sector). Finally, a sixth artefact concentration is located in the area of the loam pit (P-sector). Artefacts not attributed to one of these six concentrations are considered to belong to the remaining surface assemblage (R-sector).

In some of the concentrations, and most clearly in the C-sector (fig. 21), it is apparent that artefact registration during excavation was not sufficiently precise, resulting in a void between neighbouring excavation units.

Since there is a clear concentration of artefacts, including many refits, in the area of the loam pit, and given also the presence there of many, often large, clumps of sediment, it is probable that the present, clearly reworked, artefact concentration represents the remains of a single, original prehistoric artefact concentration in the immediate vicinity. It seems likely that this P-sector concentration originated from the nearby NE concentration.

In the NW-sector the presence of recent disturbance by a soldier's pit is clearly observable.

When analysing the refits connecting the concentrations it is obvious that the SW-sector and the C-sector have less connections than the other ones. There is no doubt that the P-sector is connected to the NE-sector, the NW-sector and the SE-sector.

The general horizontal artefact distribution (fig. 21) is essentially the same as that of the flakes (fig. 23), the blade/bladelets (fig. 24), and the chips (fig. 25). Bladelets are more scattered than blades (fig. 24), which are clearly concentrated in the NW-sector and the SE-sector. In contrast, bladelets are concentrated more significantly in the NE-sector, the C-sector and the SW-sector.

Cores and core rejuvenations (fig. 27) have clear concentrations in the NW and SE sectors. A similar distribution is characteristic for crested (fig. 28) artefacts, somewhat less so for cortical (fig. 26) artefacts. Such a distribution clearly indicates that debitage activity mainly took place at these concentrations.

Points (fig. 29) are mainly Zonhoven points. Their distribution is clearly centred on the NW-sector, but with a presence also in the SE-sector. All Ahrensburg points were found in the NW-sector. A similar observation applies to burins (fig. 30). However, it is clear that most burins are found in the NW-sector, while burin spalls are mainly concentrated in

the SE concentration, where burins are comparatively less numerous. This situation suggests that most burins were discarded in the NW-sector, but most were sharpened in the SE-sector, where they were probably most often being used. Scrapers are best represented in the C-sector and are almost entirely absent from the SE-sector, which suggests another activity zone for scrapers than for burins. Backed artefacts (fig. 32) are scattered across all concentrations. Denticulated and retouched pieces (fig. 33) are concentrated at the NW-sector and the SE-sector.

Burnt artefacts have been recorded over the whole of the excavated surface (fig. 34). They are however most numerous in the SW-sector, although there is also a relatively large number in the NE-sector. A distribution plan of the most intensively burnt artefacts provides a similar picture.

2.2 - Correlation between organic and lithic material

Charcoal fragments were rare (fig. 35) and those that were found are very small and scattered both vertically and horizontally. No charcoal concentrations that might suggest the presence of a hearth were observed. Charcoal fragments occur mainly in the NW and SW sectors, a pattern which does not entirely match the burnt artefact distribution. Indeed, in the C-sector there is a significant concentration of intensively burnt artefacts (fig. 34), yet no charcoal was found present in that concentration.

Very few of the collected samples were suitable for dating. Only two samples, both from NW-sector, have been submitted for dating. Charcoal from the SE-sector was collected in the upper part of the E horizon and thus the correlation between sample and artefact concentration was considered insufficiently close. Moreover, during a rapid inspection of these charcoal fragments E. Marinova suggested the presence of *Quercus*.

A charcoal sample (fig. 35), mainly *Pinus*, from 28.88N 7.44W (ZM94/85/59) at an elevation of 8.40 m (i.e. 0.4 m below the surface) in the podzol B2ir horizon was submitted for AMS dating and obtained a date of 7060 ± 70 BP (UtC-3195).

As the obtained result was not in accordance with the date indicated by the typology of the artefact concentration, another sample has been submitted for AMS. The second sample (ZM94/65/228), probably *Pinus*, was collected at 25.07N 5.69W at the elevation of 8.19 m (i.e. 0.6 m below the surface) in the podzol B3 horizon. The result was 10760 ± 70 BP (UtC-3720).

In the case of 25-27N 5-6W (fig. 10 and 18), where artefacts are found more than 1 m below the top of the A horizon, there is no evidence to suggest that a prehistoric pit was dug through the gravel layer and that artefacts were then dumped in it. All field evidence suggest that the absence of the cobble-gravel layer is a natural phenomenon. The significant vertical distribution of artefacts scatter in that gap is probably related to postdepositional processes, which also caused the preservation of charcoal in that area.

Both R.L.C. Atkinson (1957) and M. Armour-Chelu and P. Andrews (1994) have drawn the attention to the effects of earth worm (*Lumbricus terrestris*) burrowing on vertical artefact displacement. On the one hand, soil is brought up

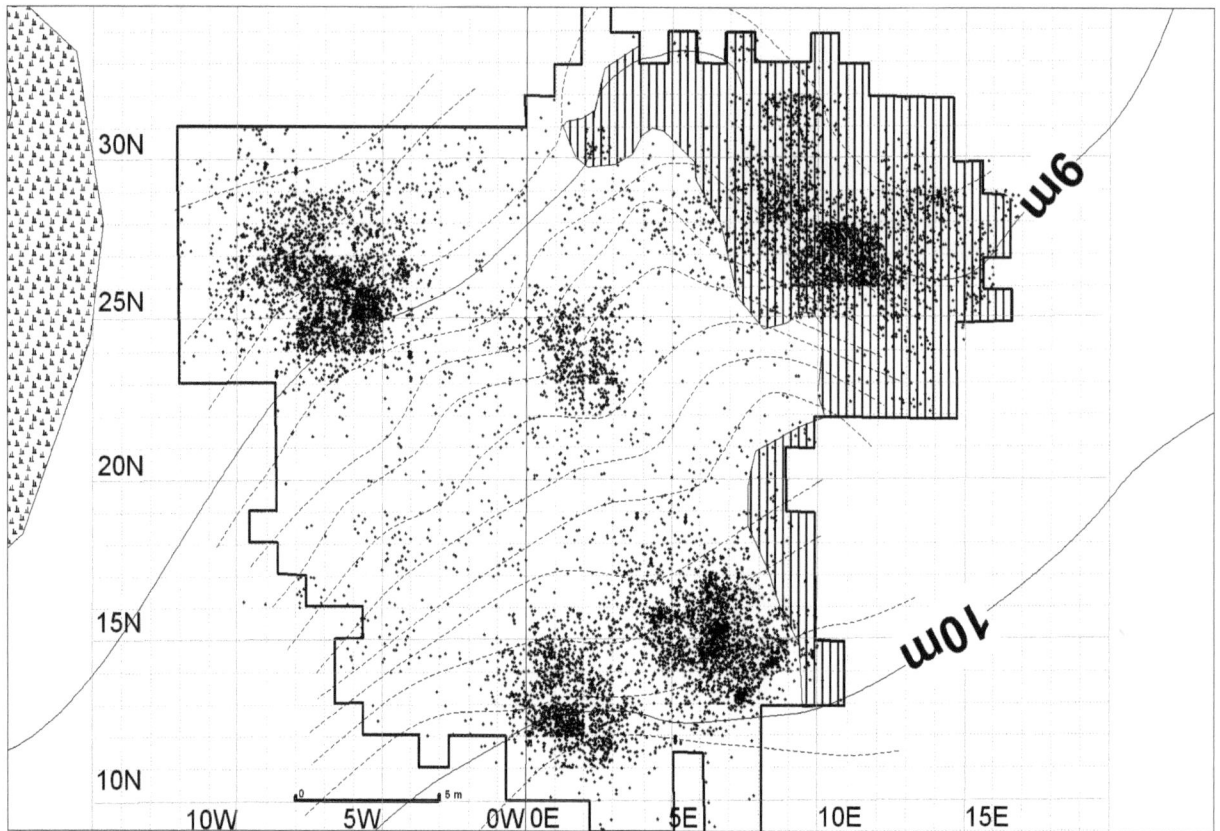

Figure 21 - Distribution of all artefacts.

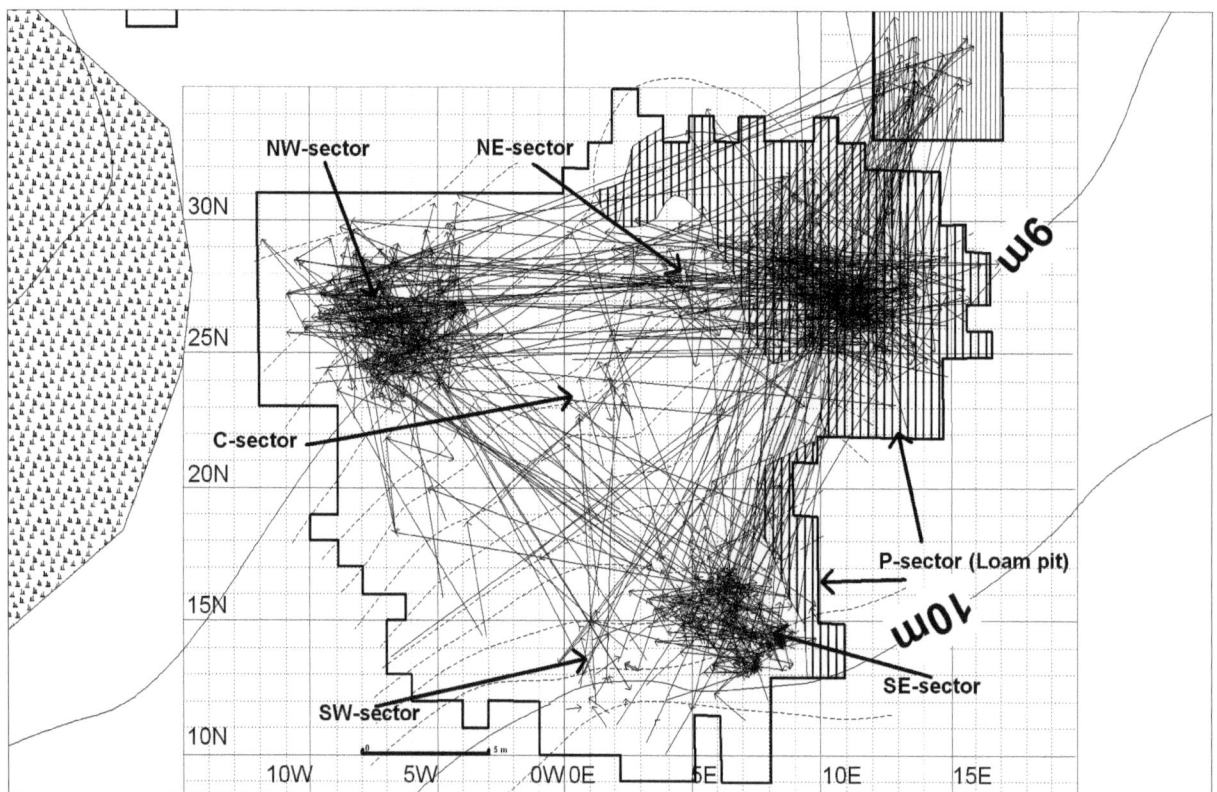

Figure 22 - Distribution of refits.

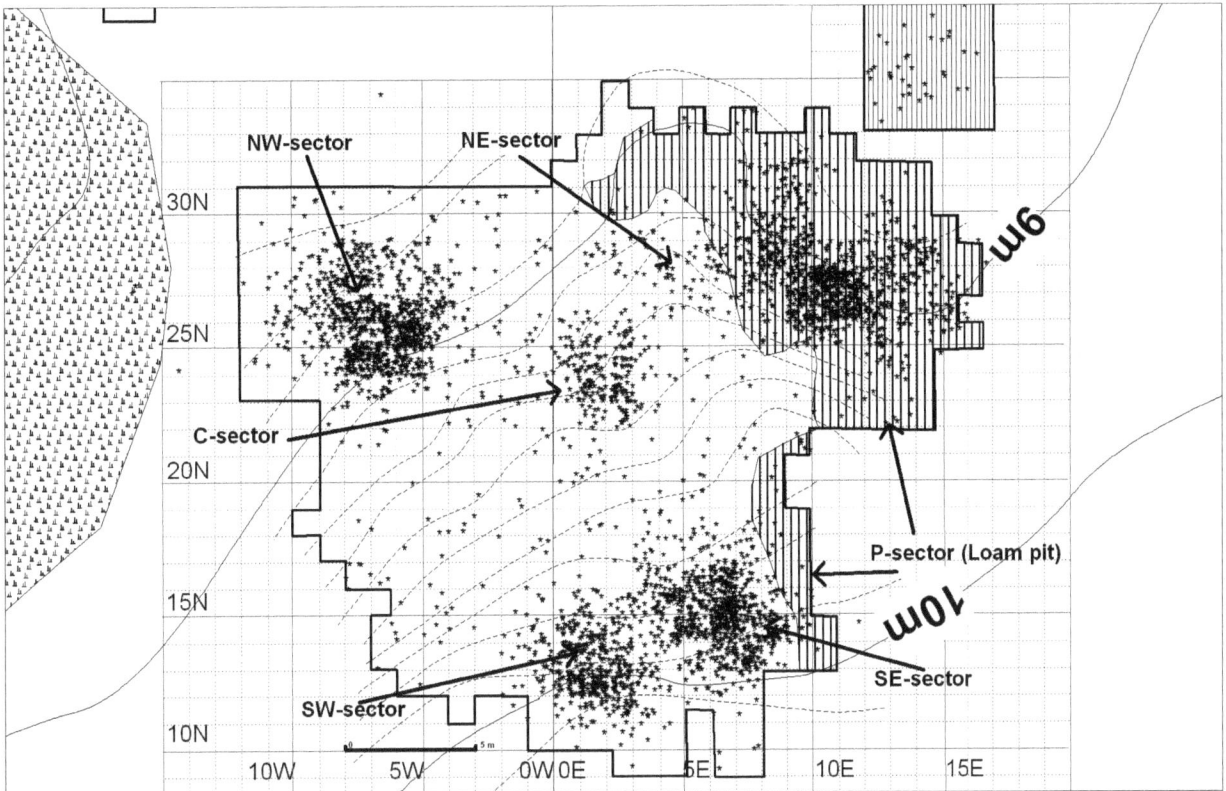

Figure 23 - Distribution of flakes.

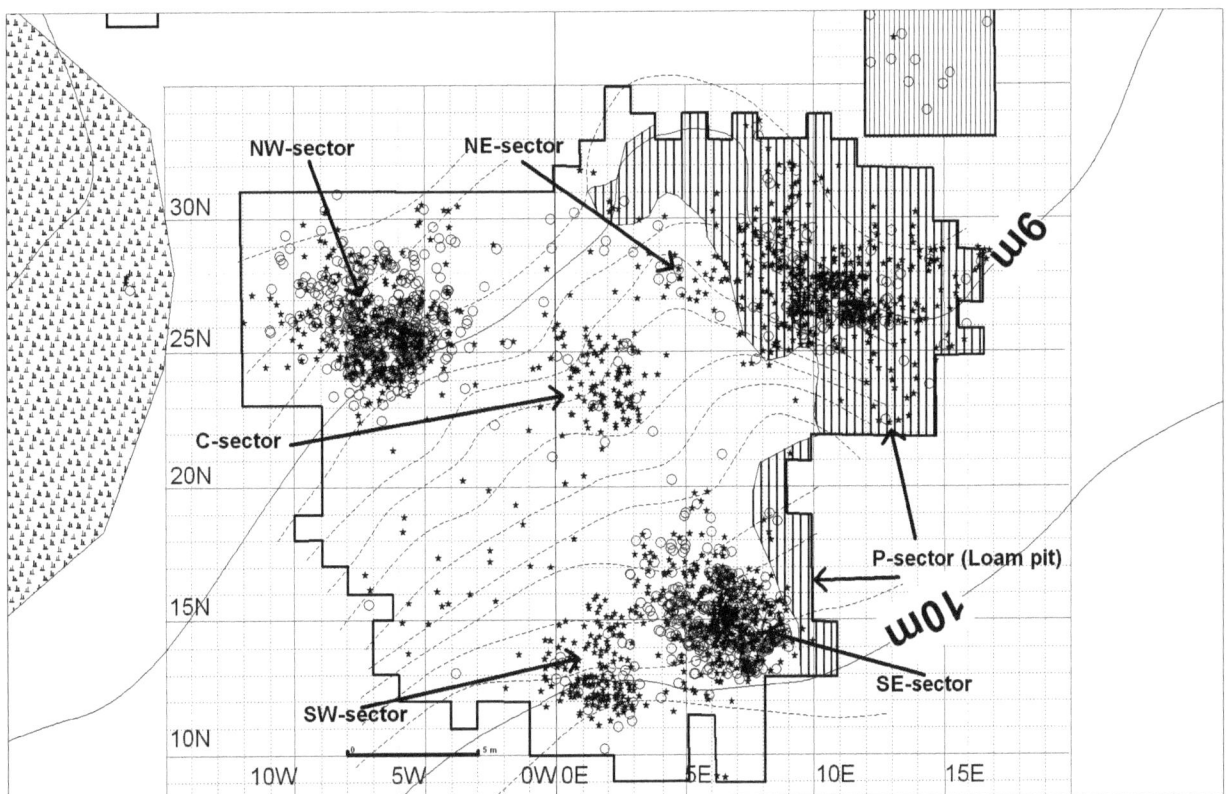

Figure 24 - Distribution of blades (circle) and bladelets (star).

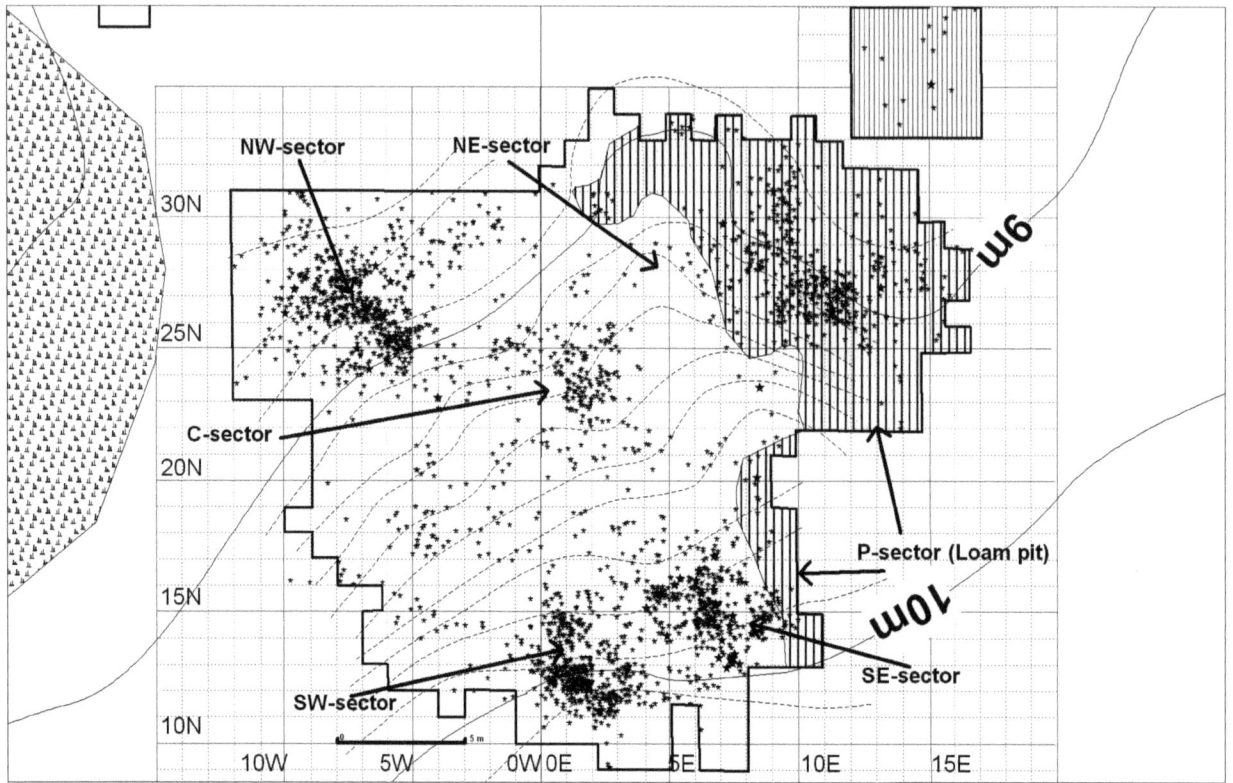

Figure 25 - Distribution of chips.

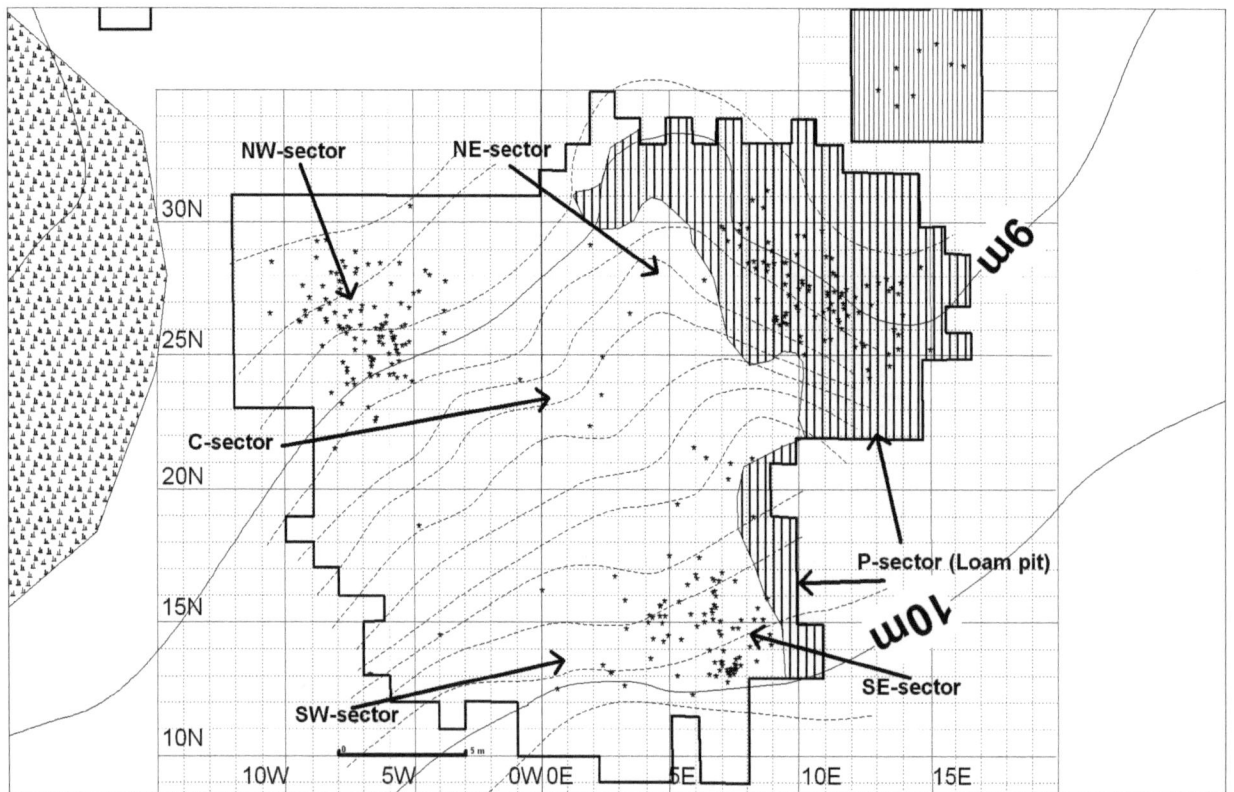

Figure 26 - Distribution of cortical artefacts.

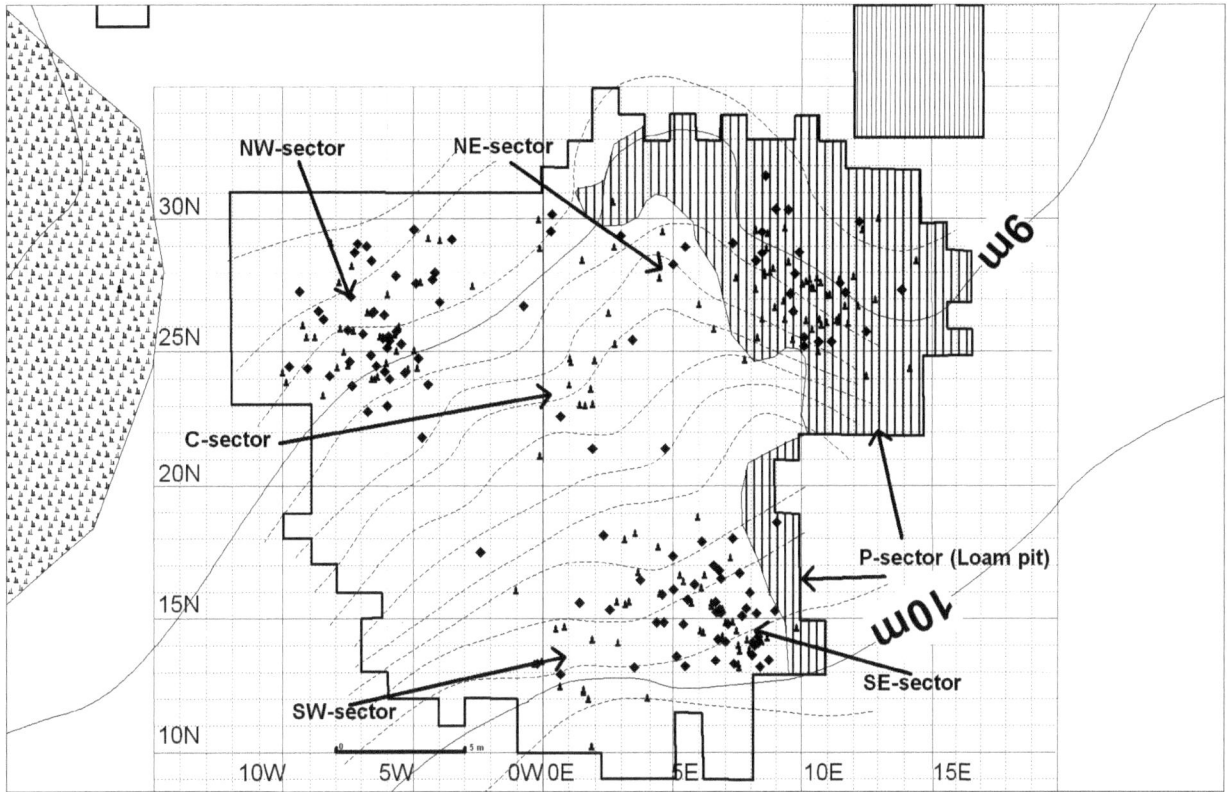

Figure 27 - Distribution of cores (rhomb) and core rejuvenations (tower).

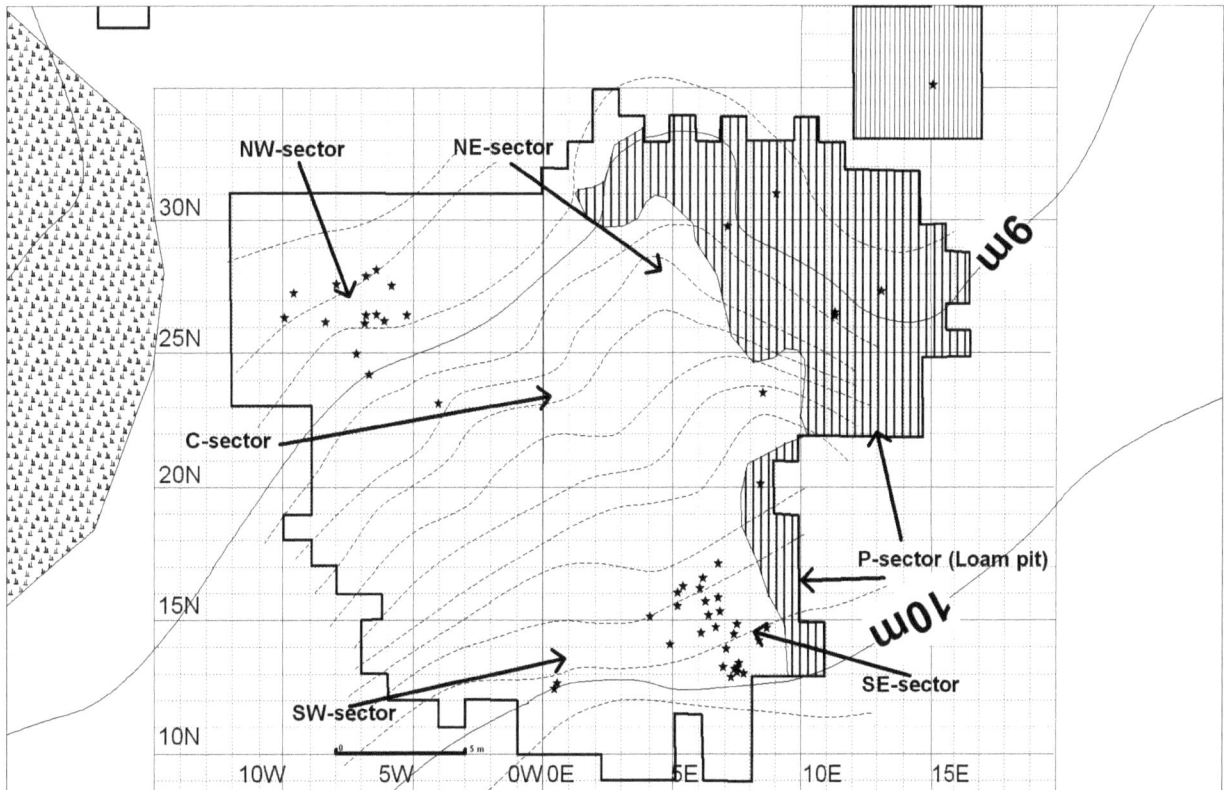

Figure 28 - Distribution of crested artefacts.

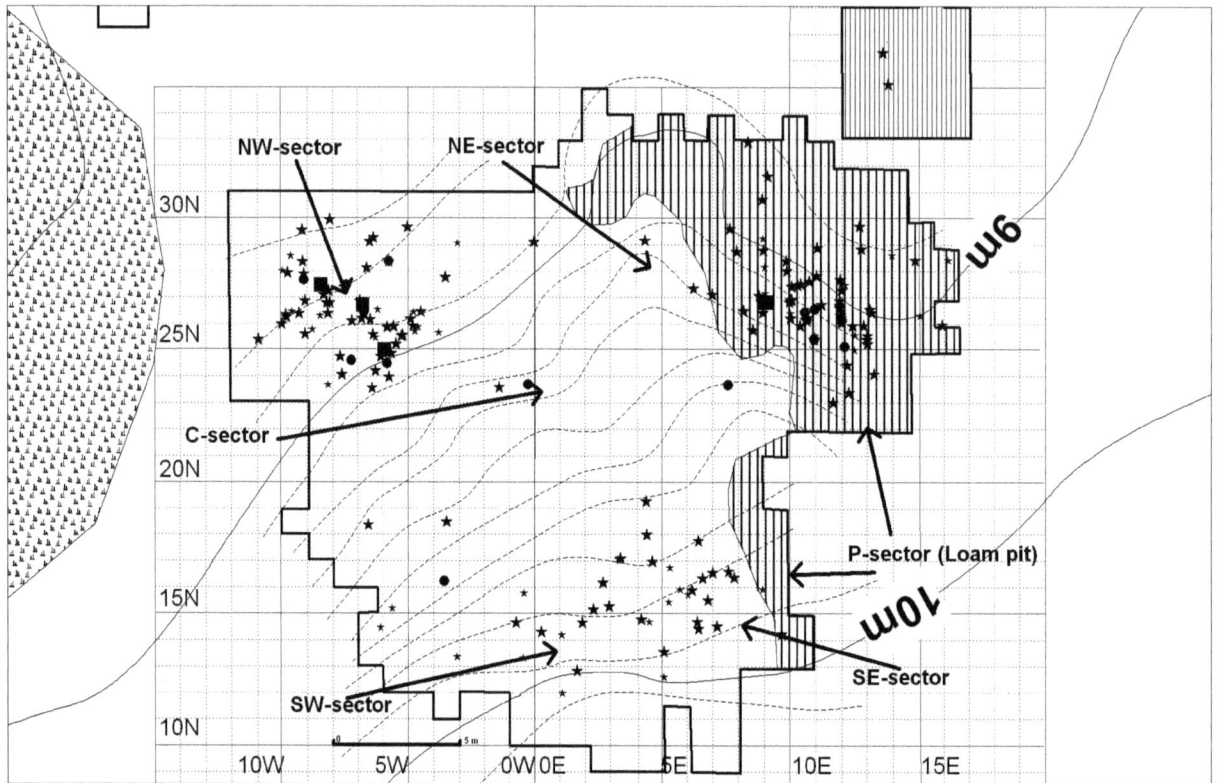

Figure 29 - Distribution of points: Ahrensburg point (black rectangle); Ahrensburg point (large star); other point (small star)).

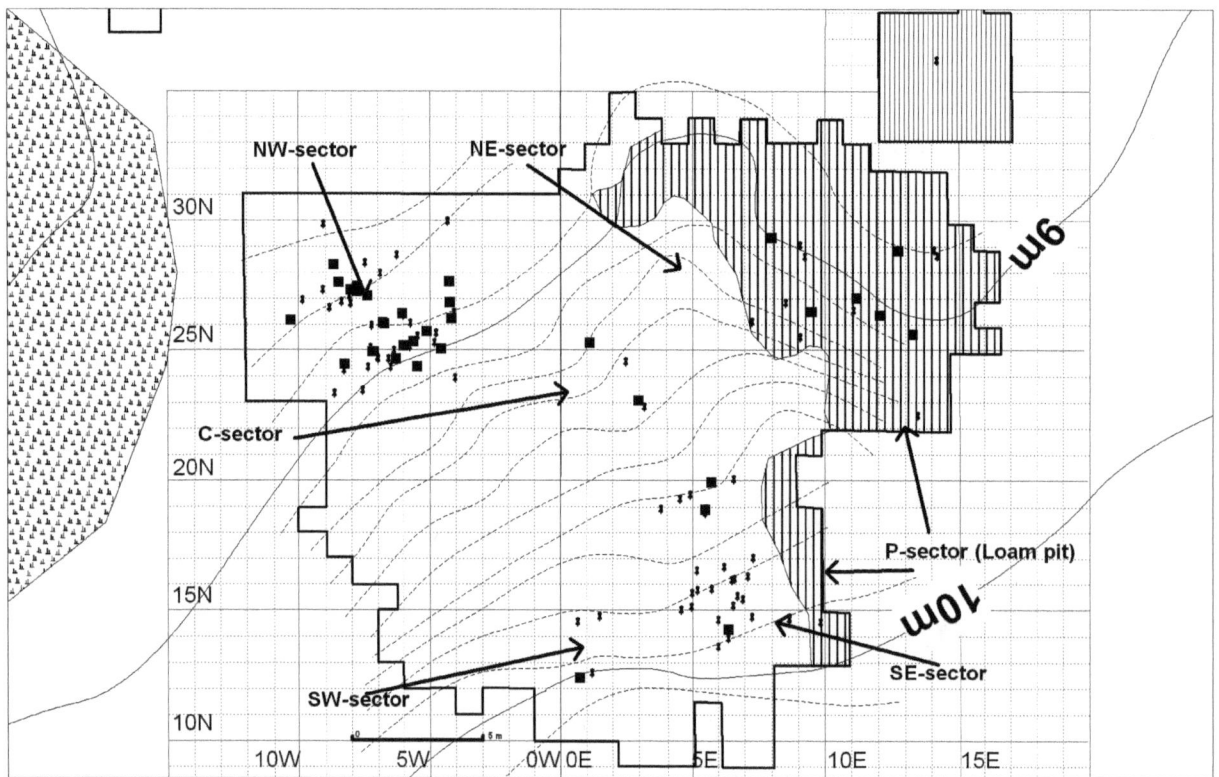

Figure 30 - Distribution of burins (squares) and burin spalls (pin).

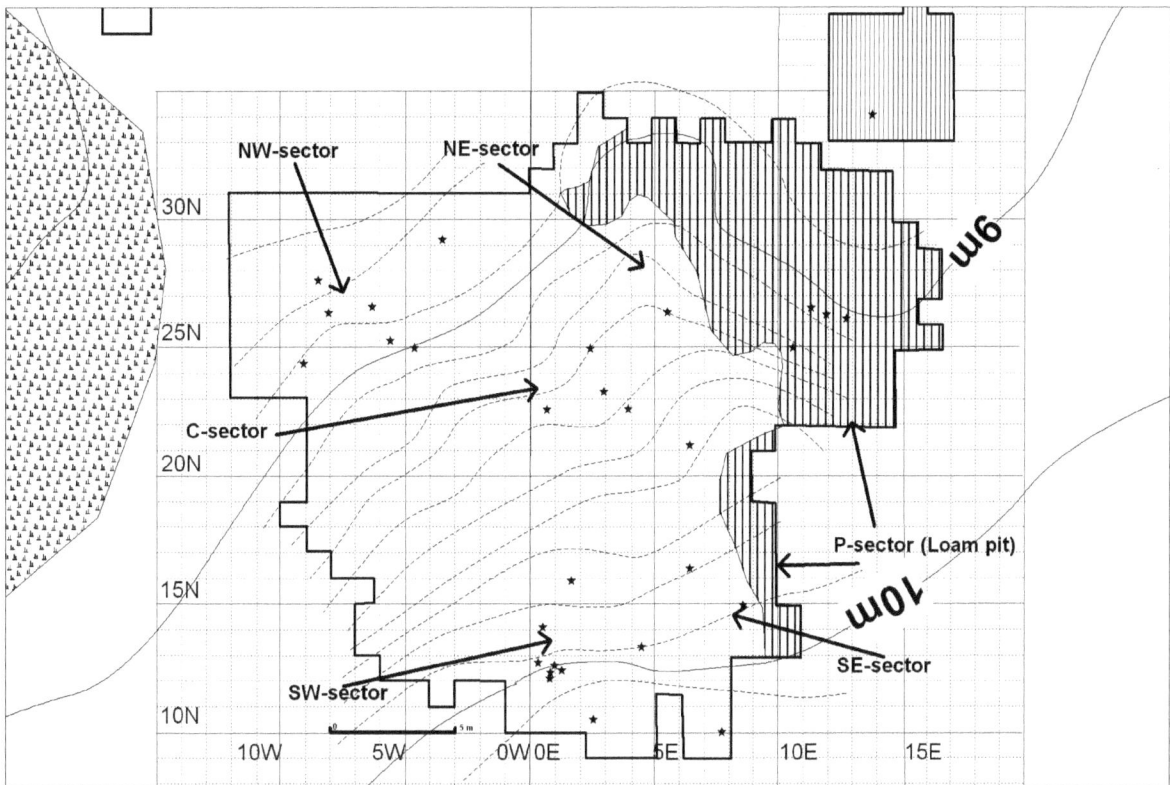

Figure 31 - Distribution of scrapers.

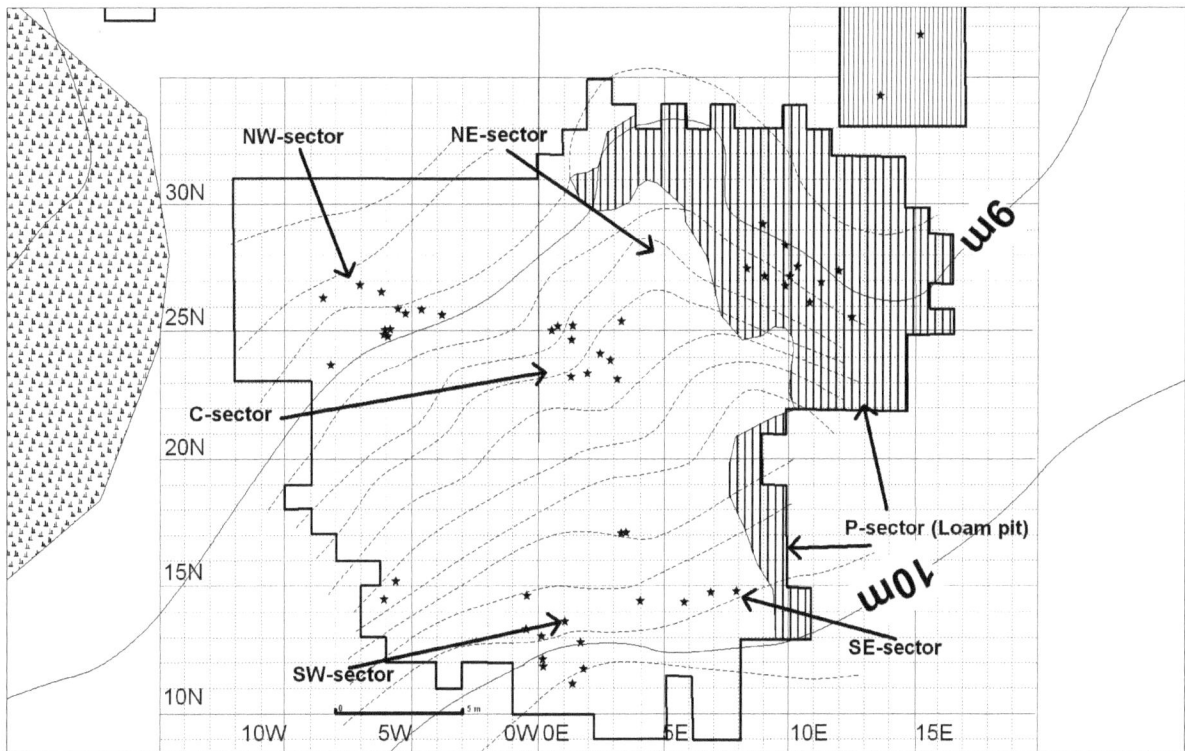

Figure 32 - Distribution of backed pieces.

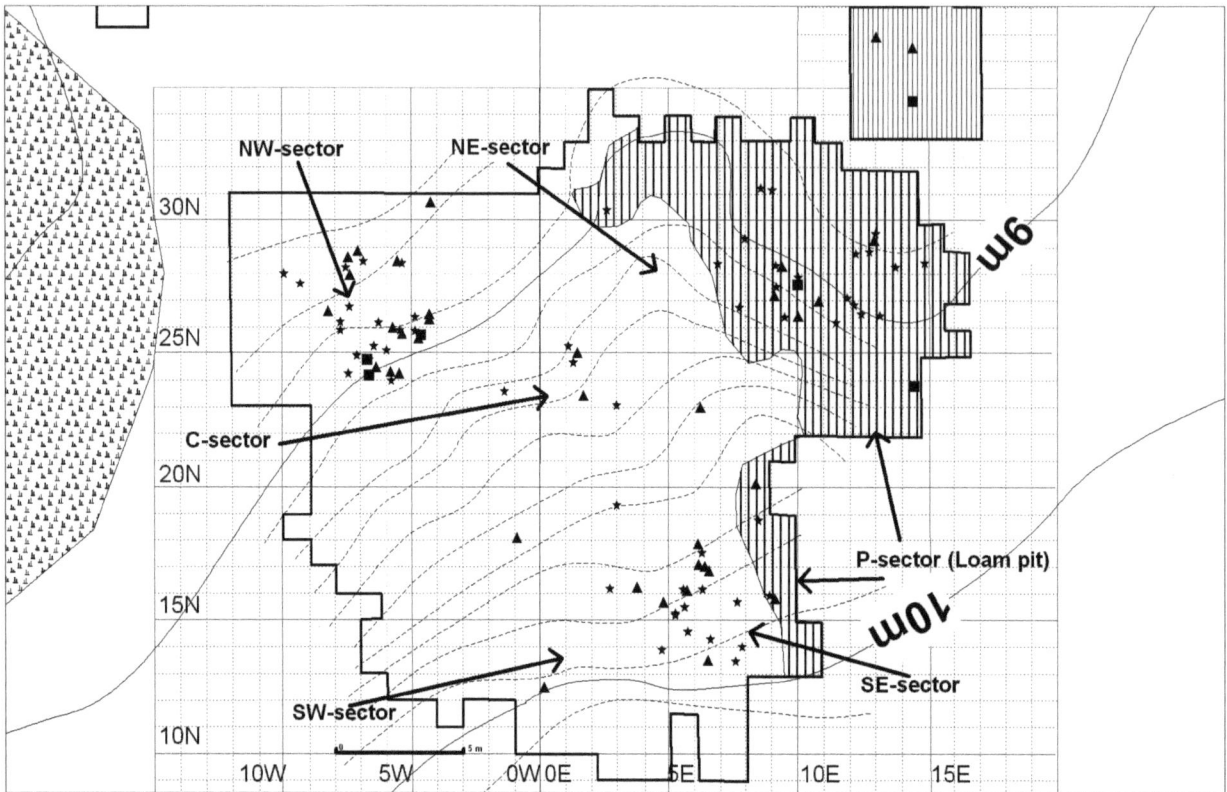

Figure 33 - Distribution of truncations (star); notched pieces (square) and retouched pieces (triangle).

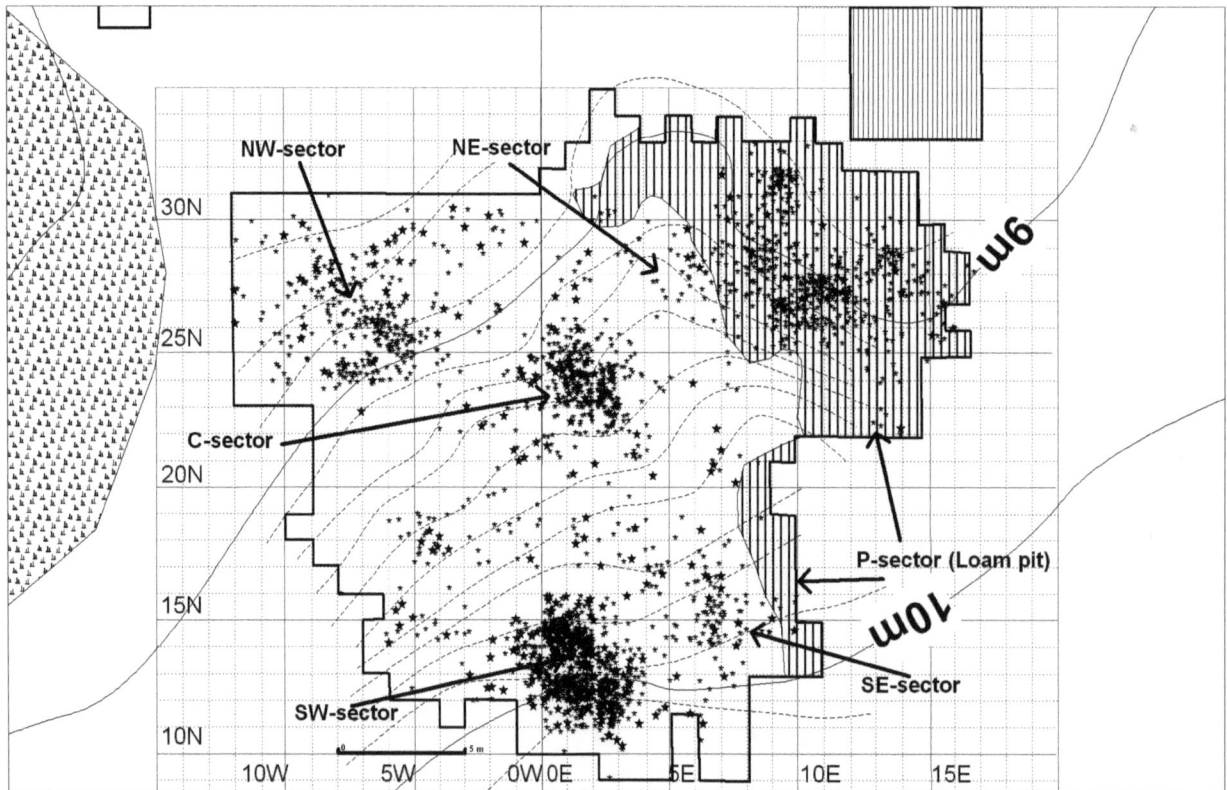

Figure 34 - Distribution of burnt artefacts, large stars indicating intensively burnt artefacts.

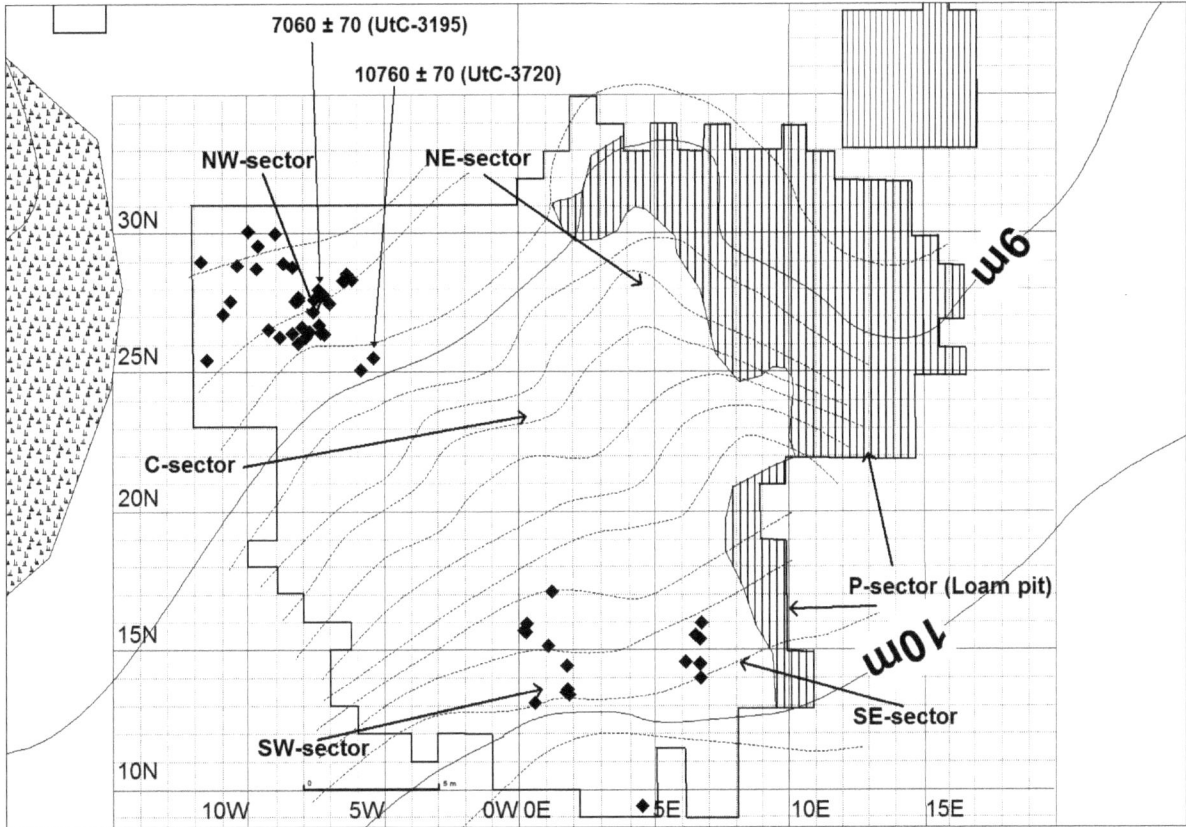

Figure 35 - Distribution of charcoal fragments showing the position of the charcoal fragments submitted for AMS dating.

to the surface as worm-casts and gradually accumulates there, while on the other, disused bur-rows below the surface are constantly collapsing, and thus producing local subsidence of the overlying soil. The net effect of these two related processes is to cause objects lying on the surface to sink below it, while the absolute level of the surface remains unchanged. The resultant rate of sinking of stones and other bodies may amount to as much as 5 mm annually (Cornwall 1958: 52). Thus where the cobble-gravel layer is absent, as in 25N 5W (fig. 17), artefacts and charcoal could have moved down deeper than in other places where the cobble-gravel layer was present and thick.

According to Stein (1983), soils with medium textures create the best habitats for earth worms as moisture is avail-

able all year round. Earthworms require an abundance of food sources. Such conditions, if not entirely optimal, were probably present at Zonhoven-Molenheide under the forest cover. It is thus quite possible that, in the forest conditions during the early Holocene, the local earth worm population was quite extensive and their activity can be held responsible for the vertical transportation of artefacts. The tolerance of earth worms to changes in acidity varies widely depending on the species involved, but most species cannot tolerate pH values below 5. As a consequence, the more recent installation of very acid conditions at Zonhoven-Molenheide and the formation of the humic-iron podzol, has made life impossible for earth worms and they have disappeared.

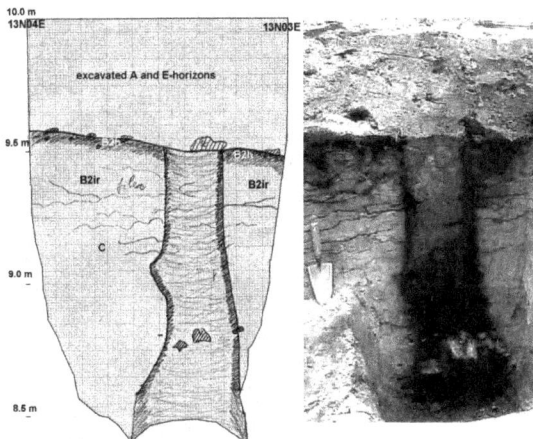

Figure 36 - Posthole 13N03E (left)

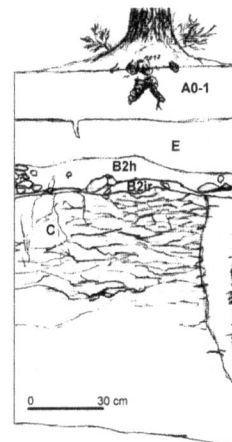

Figure 37 - Pin root at 12N5E.

We also have to take into account the burrowing activity by cockchafers and/or dung chafers (De Bakker, Edelman-Vlam 1976). The presence of their burrowing can be observed in the lower horizons where parts of the original forest soil have been preserved.

No methods to correlate charcoal and lithic material are available. The two dates available are statistically entirely different and not overlapping. There is no decisive field evidence to support the hypothesis that two different occupations left their lithics in the NW-sector. The homogeneity of the assemblage present in the NW-sector will be discussed in more detail later. Here it suffices to observe that the lithic assemblage from the NW-sector, with its numerous burins and Ahrensburg/Zonhoven points, fits best with a 10760 BP date. We therefore presume that UtC-3720 is coeval with fire-making activity at the site by the same people who left their artefacts on the surface. The charcoal of UtC-3195 is considered to be the result of a separate episode of activity in this NW-sector, which left little or no archaeological material there, but may well have done so in other contexts at the site (e.g. the artefacts in Wommersom quartzite).

2.3 - Other structures

At 12-13N7E a high concentration of 60 artefacts all of them occurring in the B2h and B2ir horizons was found (fig. 19 upper). Such a dense concentration seems to be the result of some form of intentional activity. Could it be a dump? During excavation it was observed that many artefacts were in a vertical position, yet, vertical artefact distribution was limited to a few centimetres.

A single posthole was noted in 13N03E (fig. 35). This was more than 1.3 m deep and had at its base two large cobbles. Its date remains unknown and so its relationship to the artefact deposition remains unclear

Other candidate 'postholes" were noted, but these are most probably natural products, related to the presence of pin roots from an earlier vegetation cover (fig. 33).

Several cart tracks were present (fig. 16) disturbing the Al and the upper part of the E-horizon. They are probably related to the pit dug by the loam diggers, who used the ground surface alongside the pit for transporting out the excavated loam. The breadth of the track suggests a cart with an axle width of about 1.2m.

3 - DEBITAGE

3.1 - Raw material availability

The inhabitants at Zonhoven-Molenheide exploited the flint-bearing gravel deposits, which are the remains from terrace deposits created by the river Meuse (Kempen Plateau). The proportionate presence of flint cobbles, mostly angular, is not homogeneous throughout that gravel deposit, but varies from 7 to 15% (Paulissen 1973: 219). The flint types of these gravels are generally of Maastrichtian geological origin. They are of inconsistent quality, but certainly include material appropriate for knapping purposes.

In a modern quarry, which lies some 500 m south-west of the site (fig. 3), flint cobbles are not numerous. Most often the available material is a grey flint with a rolled, angular shape. However, a survey of the quarry undertaken by the excavators in 1991 identified all flint varieties found at the site among the terrace cobbles, with the exception of black flint, type 1.

It therefore seems reasonable to suggest that the bulk of the flint material exploited at the site was recovered from such a local source. However, the exploitation of a wider region cannot be excluded. There can be no doubt that the black, finely grained flint, which is also quite well represented at the site, is not of local origin and has thus been imported. This is of excellent quality and resembles the so called Obourg flint, the precise origin of which is unknown.

Wommersom quartzite was used for a small number of artefacts (fig. 38). Most of these were collected from the loam pit (P-sector), although others were found scattered across the site. They were found near the surface or in the upper parts of the E horizon. Only two points of the total could be refitted (fig. 52, 18), one from the SW-sector in the B2h horizon and the other from the NW-sector in the E-horizon. It is difficult to determine whether all such artefacts belong to a single occupation event and thus whether the Wommersom artefacts belong with the rest of the lithic assemblage. However, most of the Wommersom quartzite artefacts are probably best considered to be intrusive and related to the occupation phase represented by the ^{14}C date of 7060 BP (UtC-3195). The two refitted points would be consistent with an assemblage of such an age.

3.2 - Flint quality

About 22 kg of flaked flint was collected from the site. If we assume that most flint nodules were collected from the terrace deposits, then it is clear that the knappers were selective in their choice from the available cobbles. The flint material from the terrace is of inconsistent quality because it has been subjected to significant weathering, principally the impact of soil formation and gelifraction. Numerous imported nodules had shattered into fragments at the site at the moment of a debitage impact. The shattered fragments were often still lying together or scattered over a small area. Other nodules appear to have broken into just a few (large) fragments which were then further reduced separately. In such instances the flat surfaces and sharp ridges of these items were advantageously exploited, as they could directly serve as striking platforms or guiding ridges. If one also takes into account the fact that the flint nodules introduced to the site were often not of usable quality, it is apparent that flint procurement was not always guided by consistent selection mechanisms.

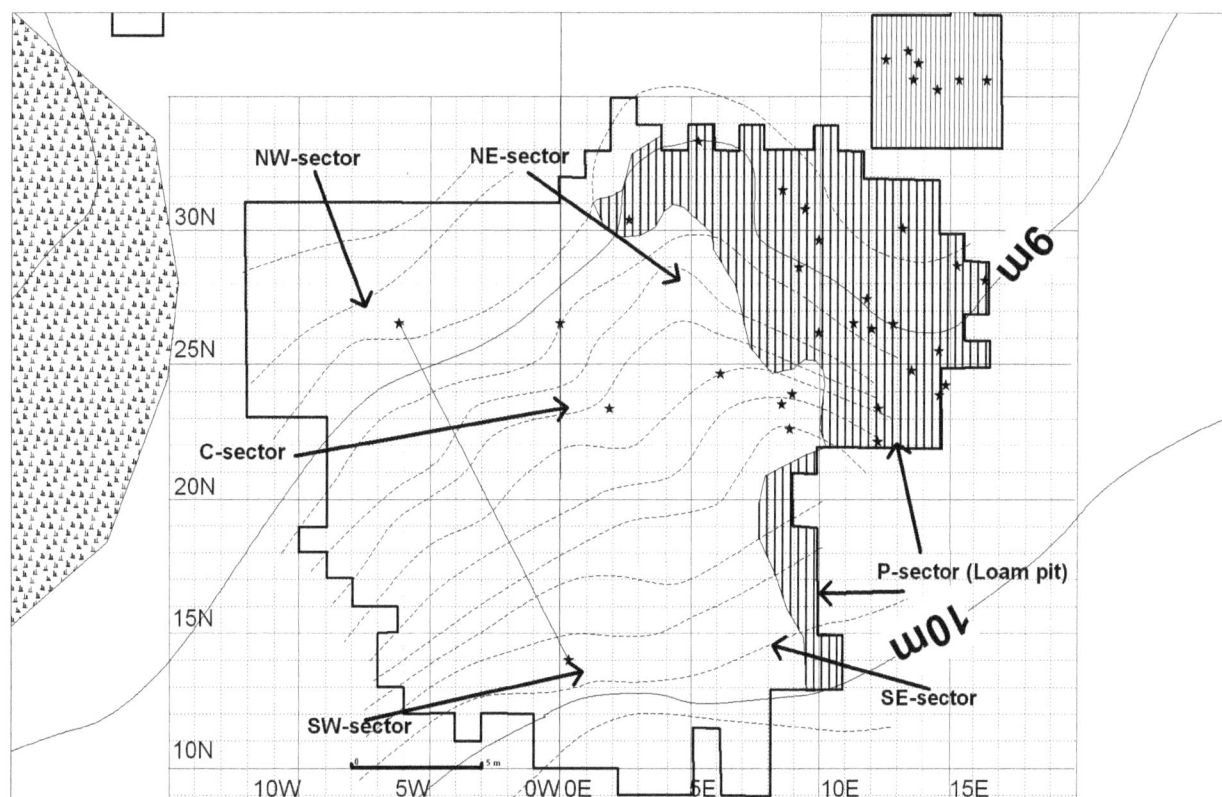

Figure 38 - Distribution of Wommersom quartzite artefacts with a single refit of two backed pointed bladelets.

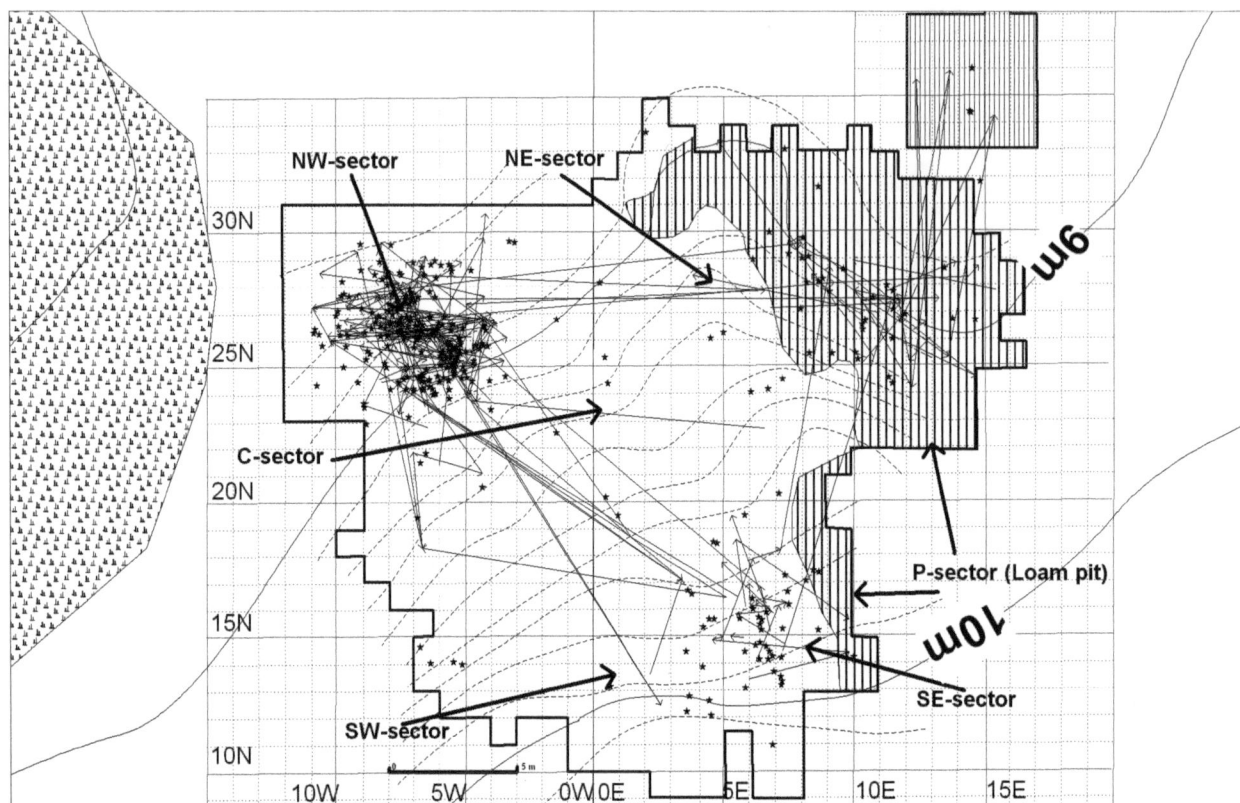

Figure 39 - Distribution of black flint artefacts and the refits between such artefacts.

3.3 - Flint types

The vast majority of the flint artefacts can be grouped into two major raw material categories: flint of diverse knapping qualities originating from the Meuse gravels (Flint type 2-9) and a black flint (Flint type 1). The distinction between these two groups is quite explicit. Meuse gravel flints are by no means an homogenous groups. Rather a range in texture, colour, and cortex variations can be observed, and this characteristic was extremely useful in the refitting exercises. This variation also permitted the definition of several, more narrowly defined flint types, based on macroscopically observed characteristics such as granulometry, colour, cortex, and the occasional presence of fossils or other inclusions.

Given the large amount of flint material, which includes many small items that could not be classified with certainty, only cores, tools, and larger blanks have so far been ascribed to a specific flint type.

The following flint qualities were observed.

Flint type 1 (black flint): Black to very dark grey, shiny flint; very homogeneous, fine-grained and of excellent flaking quality, with, when present, a white-yellow chalky cortex. Type 1 is translucent only in the very thinnest artefacts, a characteristic that differentiates it from so called Obourg flint, which is characterised by a much higher translucency.

The black flint has excellent knapping qualities. It was apparently not introduced to the site in the form of nodules collected from the Meuse gravels, but rather from an unknown source. The kind of cortex present suggests that it was probably introduced to the site in the form of a single large nodule. No large pieces were found on the site, but a minimal length for the original nodule of 16 cm is sug-

gested by refit 76. Since a total of 3.5 kg of black flint was collected on the site, this nodule should have been large. It was certainly not a fresh flint nodule. The cortex was somewhat rolled and ancient patinated frosted surfaces are present. However, interior frost fractures are certainly rare. The nodule was flaked into large blades and flakes and subsequently into bladelets. Many chips were produced and ultimately only two, fully exhausted and thinned, irregular cores were discarded. Even the cortical blanks were well adapted to tool making. Nearly all burins and burin spalls had been shaped from this black flint.

The black flint is well represented across the site, but mainly in the NW-sector, the SE-sector and P-sector (fig. 39). As such, this distribution suggests that these sectors were probably in contemporary usage

Flint type 2: Black to dark brown, fine grained flint of high quality; translucent (glassy) in thin artefacts, sometimes with small white flecks or with concentric darker zones and a white, slightly rolled cortex. This flint is often considered to originate from Obourg and is of excellent knapping quality.

Flint type 3: Dark grey medium coarse flint with rolled cortex and even a rolled cobble surface. Occasionally the presence of large white flecks may be observed.

Flint type 4: Dark grey medium fine, homogeneous flint with a slightly eroded thin white cortex and very few white speckles or more rarely small white flecks.

Flint type 5: Grey brown fine to medium grained flint with a thin slightly eroded cortex. Beneath the cortex there is a slightly darker coloured zone about 0.5 cm in thickness

Flint type 6: Grey to dark grey heterogeneous flint with black speckles.

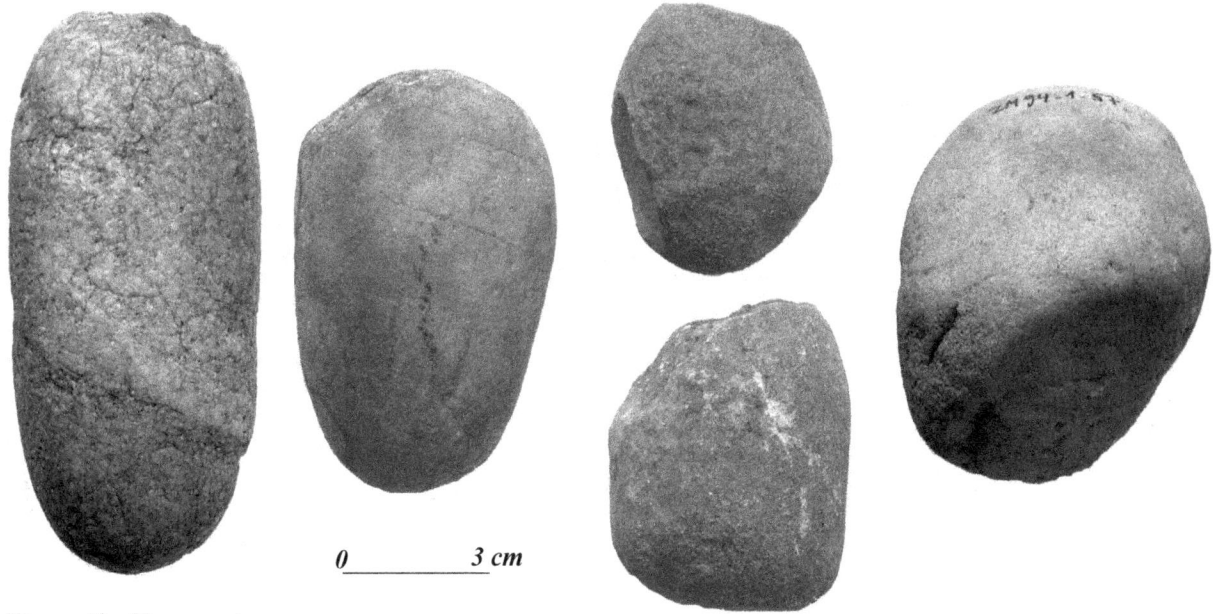

Figure 40 - Hammerstones

Table 1 - Debitage products and their characteristics; % C: % of cortical artefacts; % B: % of burnt artefacts.

Technology	Total	N	% C	% B
Blade 0% cortex	319			
Blade with < 10% cortex	153			
Blade with 10-99% cortex.	99			
Blade with 100% cortex	22			
Blade		593	46.21	6.75
Proximal blade fragm. with 0% cortex	173			
Proximal blade fragm. with < 10% cortex	70			
Proximal blade fragm. with 10-99% cortex	34			
Proximal blade fragm. with 100% cortex	13			
Proximal blade fragment		290	40.34	8.28
Distal blade fragm. with 0% cortex	64			
Distal blade fragm. with < 10% cortex.	30			
Distal blade fragm. with 10-99% cortex	18			
Distal blade fragm. with 100% cortex	6			
Distal blade fragment		118	45.76	9.32
Medial blade fragm. with 0% cortex	43			
Media blade fragm. with < 10% cortex	14			
Media blade fragm. with 10-99% cortex	10			
Medial blade fragm. with 100% cortex	2			
Medial blade fragment		69	37.68	11.59
Bladelet 0% cortex	730			
Bladelet with < 10% cortex	165			
Bladelet with 10-99% cortex	99			
Bladelet with 100% cortex	23			
Bladelet		1017	28.22	9.05
Proximal bladelet fragm. with 0% cortex	223			
Proximal bladelet fragm. with < 10% cort.	53			
Proximal bladelet fragm. with 10-99% cor.	30			
Proximal bladelet fragm. with 100% cortex	1			
Proximal bladelet fragment		307	27.36	20.85

Technology	Total	N	% C	% B
Distal bladelet fragm. with 0% cortex	144			
Distal bladelet fragm. with < 10% cortex	39			
Distal bladelet fragm. with 10-99% cortex	37			
Distal bladelet fragm. with 100% cortex	5			
Distal bladelet fragment		225	36.00	12.00
Medial bladelet fragm. with 0% cortex	74			
Medial bladelet fragm. with < 10% cortex	25			
Media bladelet fragm. with 10-99% cortex	22			
Medial bladelet fragm. with 100% cortex	1			
Medial bladelet fragment		122	39.34	12.30
Flake 0% cortex	748			
Flake with < 10% cortex	322			
Flake with 10-99% cortex	195			
Flake with 100% cortex	89			
Flake		1354	44.76	10.78
Flake fragm. with 0% cortex	1091			
Flake fragm. with < 10% cortex	412			
Flake fragm. with 10-99% cortex	200			
Flake fragm. with 100% cortex	157			
Flake fragment		1860	41.34	30.91
Total blades, bladelets and flakes	5955	5955	39.40	16.83

Flint type 7: Light grey coarse flint with slightly rolled cortex and a 1 cm thick brown zone beneath the cortex. It might be related to the Valkenburg flint.

Flint type 8: Light grey medium fine mainly homogeneous flint with rare white flecks. This flint has fine flaking properties and is traditionally called 'Hesbaye Flint'.

Flint type 9: Light grey medium grained flint, with fine white speckles, similar in colour and texture to Wommersom quartzite. It presents a white slightly rolled cortex.

Flint type 10: Whitish grey fine nearly homogenous flint with rare black stripes and a slightly eroded white, thin, cortex.

Table 2 - General inventory of the artefacts by sector.

	SW	%	SE	%	NE	%	NW	%	C	%	P	%	Rest	%	Total	%
	1	0.06	11	0.47	3	2.56	9	0.33	1	0.13	5	0.02	0	0.00	30	0.26
core pyramidal	1	0.06	1	0.04	0	0.00	3	0.11	0	0.00	0	0.00	0	0.00	5	0.04
core opposed platform	2	0.13	13	0.55	0	0.00	16	0.59	1	0.13	13	0.45	0	0.00	45	0.39
core crossed platform	0	0.00	5	0.21	0	0.00	3	0.11	0	0.00	0	0.00	1	0.08	9	0.08
core irregular	4	0.26	10	0.43	0	0.00	4	0.15	1	0.13	1	0.03	1	0.08	21	0.18
blade	33	2.14	253	10.77	8	6.50	357	13.24	14	1.84	91	3.18	41	3.13	797	6.84
blade fragment	16	1.04	135	5.74	4	3.25	88	3.26	7	0.92	165	5.76	16	1.22	431	3.70
bladelet	150	9.73	389	16.55	22	17.89	308	11.42	101	13.27	467	16.30	116	8.85	1553	13.33
bladelet fragment	10	0.65	29	1.23	2	1.63	10	0.37	6	0.79	22	0.77	26	1.98	105	0.90
flake	488	31.65	786	33.45	47	38.21	858	31.81	243	31.93	942	32.88	218	16.64	3582	30.75
flake fragment	25	1.62	21	0.89	5	4.07	84	3.11	9	1.18	91	3.18	59	4.50	294	2.52
core rejuvenation	13	0.84	50	2.13	4	3.25	43	1.59	11	1.45	45	1.57	5	0.38	171	1.47
chip	537	34.82	402	17.11	13	10.57	586	21.73	203	26.68	472	16.47	589	44.96	2802	24.06
chunk	236	15.30	211	8.98	10	8.13	196	7.27	143	18.79	430	15.01	219	16.72	1445	12.41
Krukowski microburin ??	0	0.00	0	0.00	0	0.00	1	0.04	0	0.00	0	0.00	0	0.00	1	0.01
burin spall	4	0.26	20	0.85	2	1.63	33	1.22	2	0.26	12	0.42	1	0.08	74	0.64
Siret flake	2	0.13	2	0.09		0.00		0.00		0.00	5	0.17	0	0.00	9	0.08
tool	28	1.82	52	2.21	6	4.88	133	4.93	22	2.89	123	4.29	20	1.53	384	3.30
Total	1542	100.00	2350	100.00	123	100.00	2697	100.00	761	100.00	2865	100.00	1310	100.00	11648	100.00

Aside from flint type 1, all other flint types (about 18.5 kg) seem to originate from the frosted Meuse cobbles.

3.4 - Nodule Morphology

There were apparently no specific morphological preferences. Several general types may be distinguished, such as nodules that are spherical, sub-cylindrical, egg-shaped or knobbed (i.e. irregularly shaped and covered with salient protuberances, deep cavities, and angular nodules). Although, in terms of size, some preference may have existed for fist-sized nodules, flint procurement did not employ particularly strict selection criteria

The cortex when present is often extremely worn or is even sometimes smooth. More rarely nodules are coated with a chalky cortex. Sometimes the collected nodules present a (naturally) flaked surface which exhibits a whitish patina.

In conclusion, the flint knappers at Zonhoven Molenheide exploited a wide range of raw material, most of which seems to be of local origin.

3.5 - Debitage products

A detailed inventory of debitage products is provided in table 1, while an inventory by sector may be found in table 2.

One, rather surprising feature of the assemblage is that flakes are more numerous than chips (flakes less than 20 mm long). Normally such chips are by far the best represented artefact type at sites where knapping took place. This pattern can be explained as a product of the excavation method. Owing to the very high number of frosted flint chips present, it was too time-consuming to retrieve all chips from the sieve and this inevitably meant that actual, human produced chips were also disregarded. These tiny elements do not provide extensive technological information, and did not play a role in the refitting process. Their analytical value lies mainly in their spatial patterning, which, in any case, is not different from that of the flakes (fig. 23 and 25). Of course, as argues earlier, wind

deflation may also have played a role.

In total the lithic assemblage from Zonhoven-Molenheide consists of some 11,342 items. At the time of writing, 1,697 (14.4%) of the artefacts larger than 2,0 mm could afflictively be included in one of the 548 refitting groups. The refits are presented in detail in section 4.4.2.

Flakes and blades

In total, 3,722 flakes and fragmented flakes and 2,771 blades, bladelets and fragments (i.e. a total of 6,493 unmodified blanks > 20mm) were recorded.

Flakes are generally not large, most being smaller than 5 cm, excepting again those of the black flint and those obtained from frosted nodules. Cortical artefacts (39%) are numerous (tab. 1). The predominance of cortical flakes in this category is the result of the presence of numerous flakes originating from frosted nodules which were rapidly abandoned because of their boor flaking properties.

The technological features of blade(let)s and flakes, including overall dimensions, types of butt, descriptions of bulbs of percussion, percussion ripples, blade profiles, overhang preparation, flaking angles, will not be considered here in great detail.

It would appear that nodules were directly exploited by a laminar reduction whenever possible. In a random sample the large majority of both blade(let)s and flakes carry a plain (or flat) butt. Other butt types, such as cortical, dihedral, facetted or linear, also occur, but are far less common. Spurs ('talon en éperon') are completely lacking. In fact, there are no butt types specific to blades, rather facetted and dihedral types also occur on flakes.

Percussion marks (impact points of the hammer) on the butts are absent in the majority of cases. However, they are present on about one third of the butts, mainly faintly preserved, but sometimes more clearly marked. Examples of detached butts are also present. Distinctions between flakes and blades with respect to percussion marks are

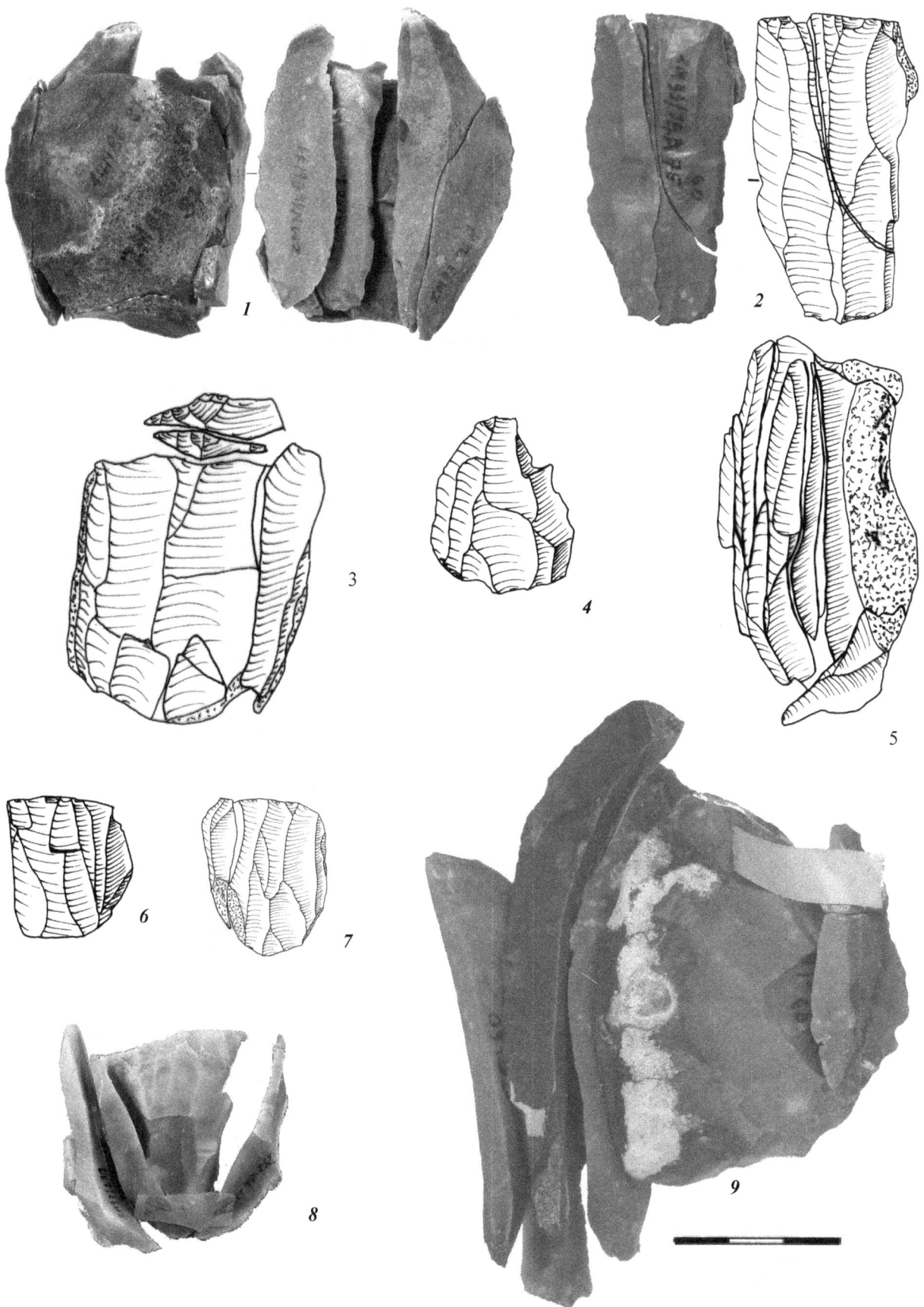

Figure 41 - Cores: 1-5, 7, 9: opposed platform core; 6: single platform core; 8: pyramidal core (from NW: 3, 5; from SW: 7; from SE: 1-3, 9; from P: 1, 4, 6, 8).

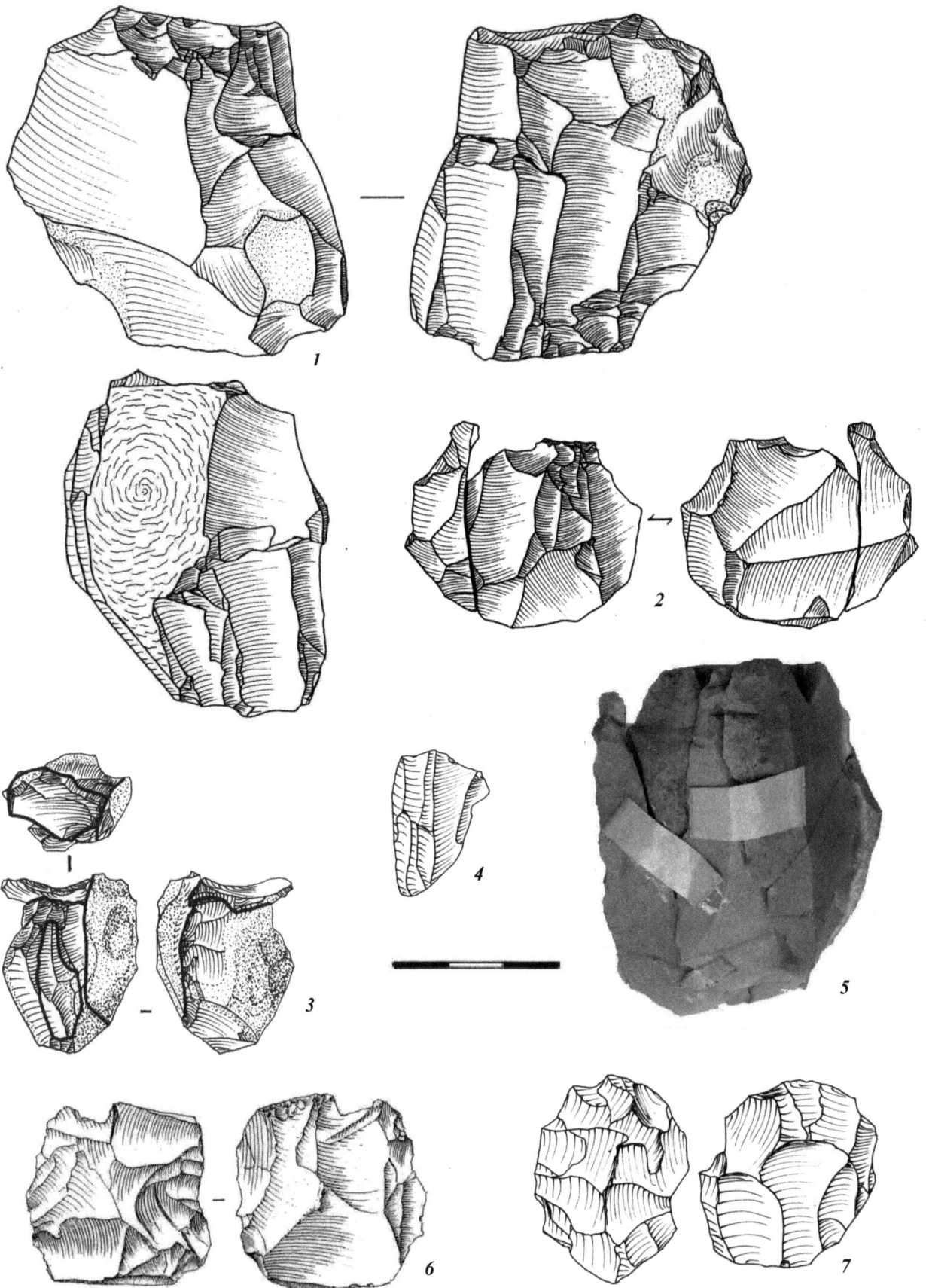

Figure 42 - Cores: 1-3, 6: opposed platform core; 4: pyramidal core; 5: single platform core; 7: crossed platform core (from NW: 3-4, 6; from NE: 2; from R: 1, 7; from P: 5).

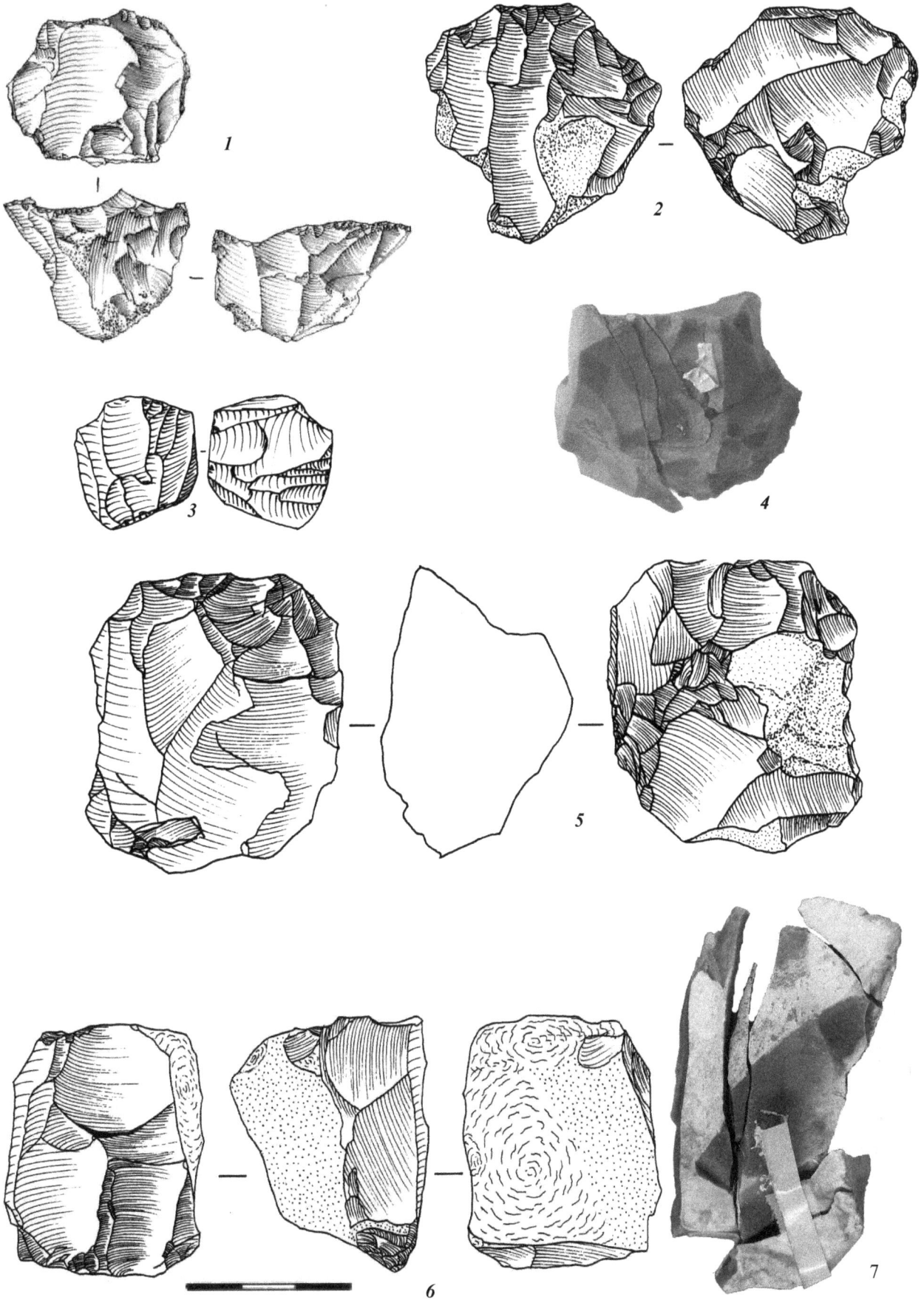

Figure 43 - Cores: 1: pyramidal core; 2, 5-7: irregular core; 3: crossed platform core; 4: single platform core (from NW: 1, 3, 7; from SE: 5, 6; from R: 2,4).

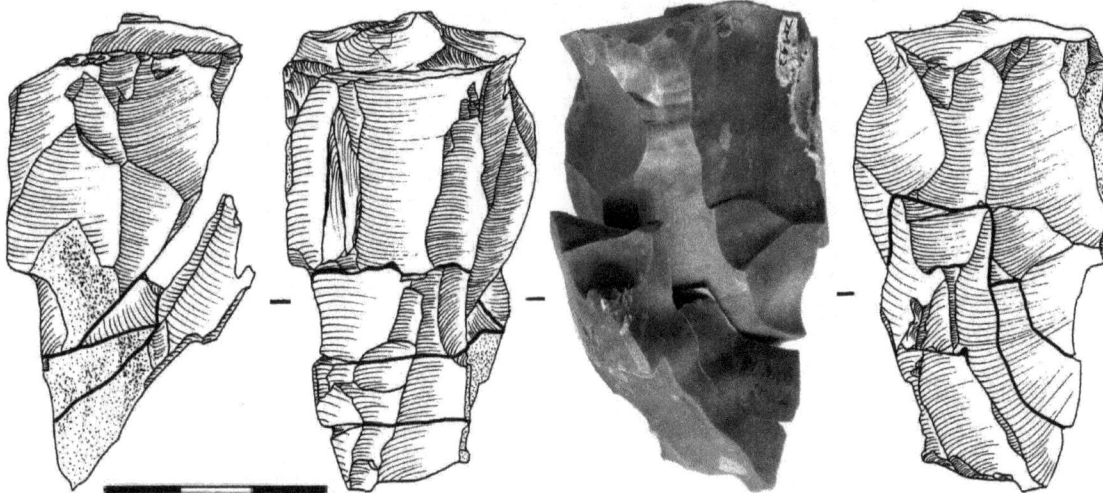

Figure 44 - Opposed platform core: Refit 15 from SE.

minimal, suggesting that hammer types were not fundamentally different for the production of either type of blank. However, further analysis on this topic is certainly required (De Bie 1999).

The bulbs of percussion on the blades are often slightly marked, but pronounced salient bulbs also occur. In profile, the blades and bladelets are more often curved than straight. Bulbs on flakes are clearly more substantial.

The flint knapping at Zonhoven-Molenheide was largely dominated by the use of direct hard hammer percussion throughout the reduction process.

Some Siret accidents are present but not numerous. A single piece could possibly be a Krukowski microburin. Microburins are entirely missing.

Hammer stones were present (fig. 40) but it is rather difficult to distinguish real hammer stones from cobbles with impact traces.

Core rejuvenation products

The 356 recognised core rejuvenation products comprise crested pieces, tabular flakes, and core sides. The distinction between these categories is not always clear-cut.

Crested pieces

Most of the complete crested pieces have laminar dimensions (fig. 46, 3, 8-10). Many of them are fragmented. The vast majority of these artefacts are unidirectional crests and in a rather rudimentary style, but some bidirectional crests are present. The number of removal negatives is generally small, and the area where they occur on the dorsal face is often limited.

Most of the tabular flakes seem to have removed only part of the core's striking platform. Real core tablets, that is pieces detaching the whole platform, are rare.

Core rejuvenations

In all, 129 rejuvenation elements have been listed as core sides, that is pieces which have removed all or a large part of the core's reduction face.

If available, natural surfaces were used as the striking plat-

form without preparation. In the course of the reduction sequence, these might be modified once or twice, but never more than three times. While single and multiple platform cores exist, cores with double platform cores are dominant, with most often an alternating exploitation of the opposed platforms. The type and intensity of core preparation did not necessarily improve the quality of the out-put. Some of the better blade sequences were obtained from poorly prepared nodules and vice versa. In other words, there is no direct correlation between the quality of the preparation and the results in terms of productivity.

Cores

Most cores are obtained from Meuse cobble flint; only 3 specimens are of the black flint type 1.

In this inventory of 98 cores the so-called tested nodules are not included. There are also many frosted nodules that were imported into the site but had scattered at the first blow. Only nodules with at least three traces of flaking have been included.

The degree of standardisation in the core assemblage is very low. Most cores are quite exhausted. In the good quality black flint type 1, exhausted cores tend to obtain rather flat volumes because at the end flaking took place on both sides of the core (fig. 41,4; 42, 7). The opposed platform cores (fig 41, 1-5, 7-9; 42: 1, 3-4; 44) are most numerous. This seems to be the normal flaking approach on the site. Single platform (fig. 41, 6; 42, 5; 43, 4) and pyramidal (fig. 41, 7, 9; 42, 4; 43, 1) cores often relate to frosted nodules that had only a restricted use. Crossed platform cores (fig. 42, 2, 7; 43, 3) are cores that have been initiated as opposed platform cores but at the point of exhaustion the knapper saw the possibility of using the core table as a new platform for producing a limited number of artefacts.

Finally, the shape of 21 cores could only be qualified as irregular (fig. 43, 2, 5-7). Here the ultimate shape of the core was essentially determined by the natural morphology of a frosted flint nodule.

3.6 - Knapping products

Overall it would seem that knapping was principally directed at the production of short blades and bladelets, the exception again being the black flint which was used for the production of long blades. It seems that the debitage

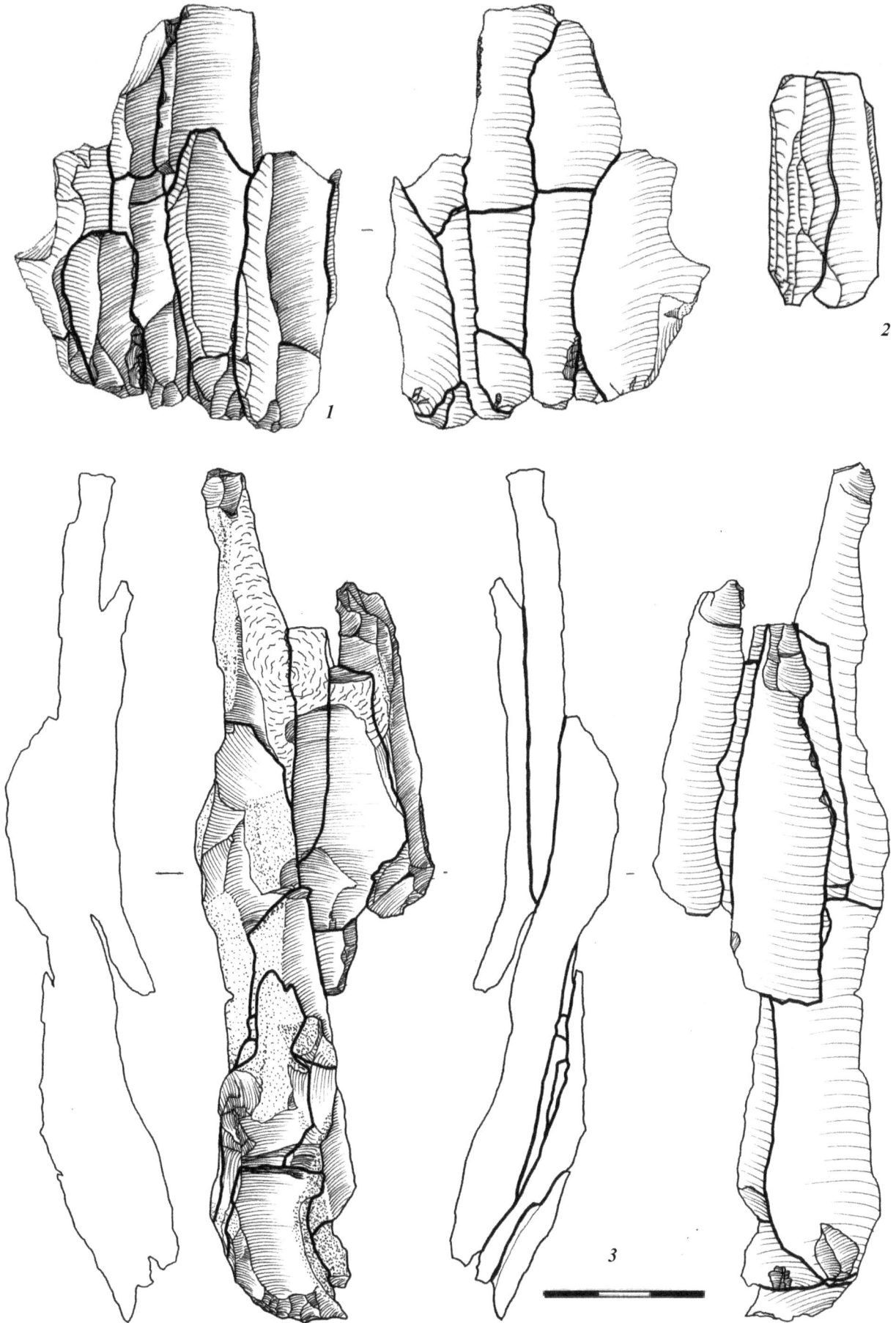

Figure 45 - Refitted blades from opposed platform cores (1: refit 11 from P; 2 from R; 3: refit 76).

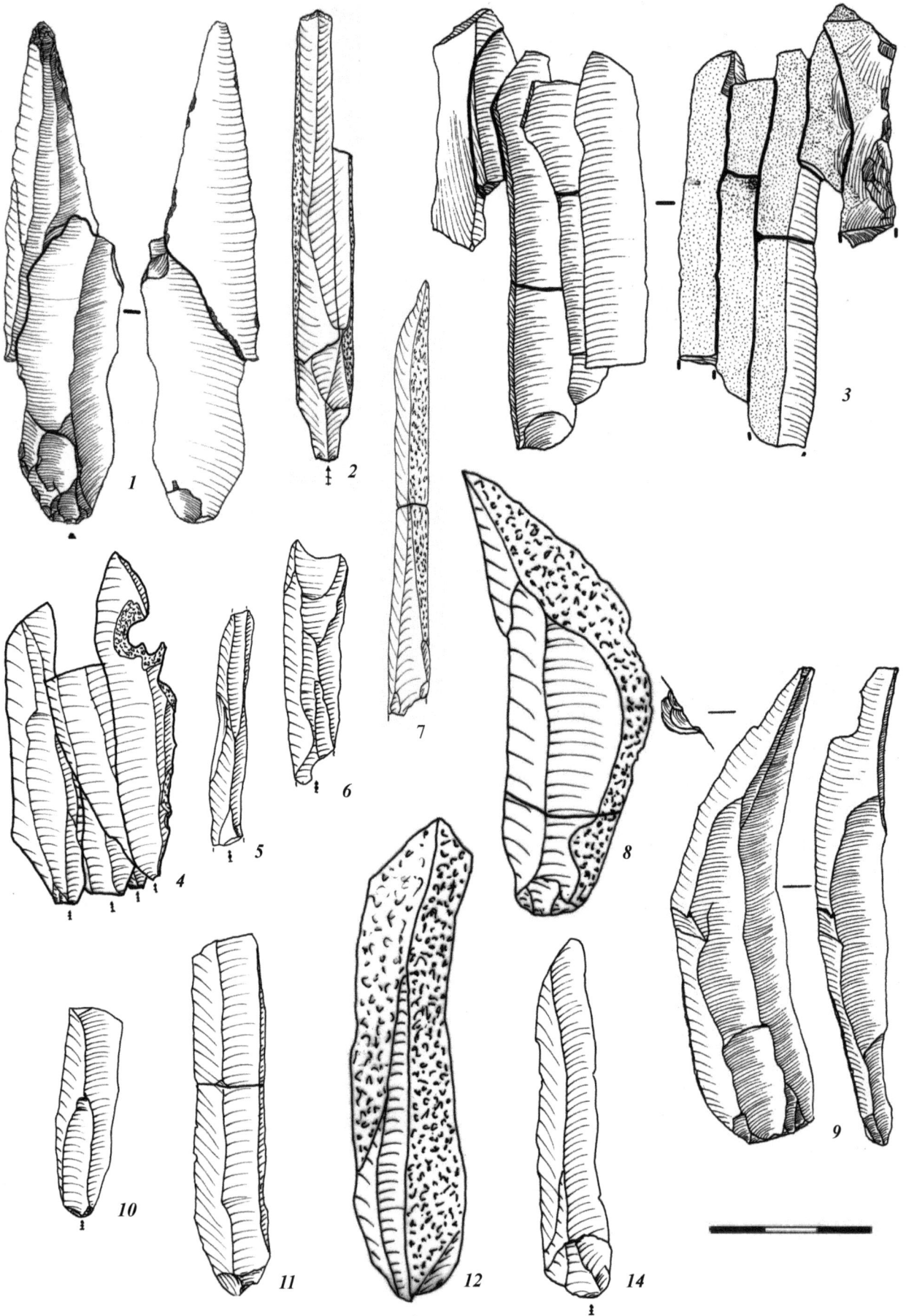

Figure 46 - Blades and bladelets (1: refit 20 from SE; 2: refit 120 from SE and NW; 3: refit 199 from SE; 4: refit 12 from NW and P; 5: from R; 6 from NW; 7: refit 202 from SE; 8: from SE; 9 from R: 10: from SE; 11: refit 143; 2, 12: from SE; 13: refit 143b from R; 14: refit 143c from P).

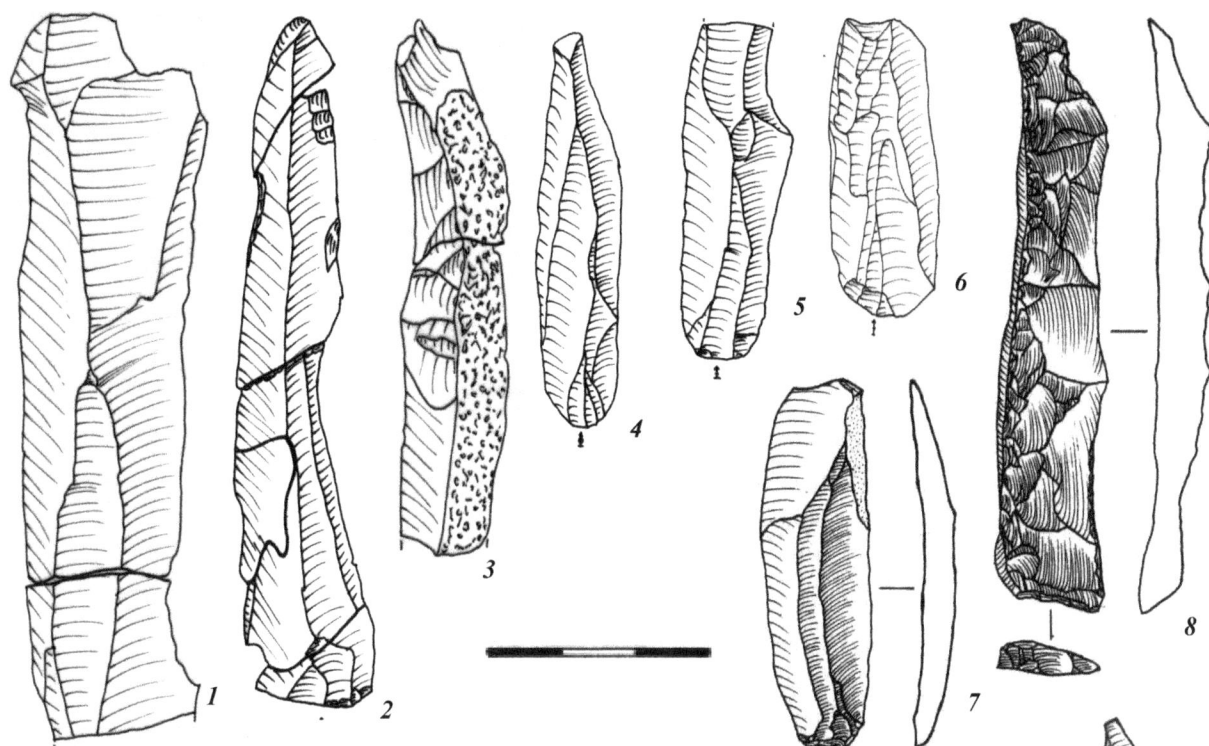

Figure 47 - 1-2, 4-7: blades; 3, 8-10: crested blades (1-2, 6: from NW; 3-4: from SE; 45: from R; 8: from R; 9-10: from P).

Figure 48 - Refit 361: an important series of flakes and blades from an opposed platform core which is missing from the refit.

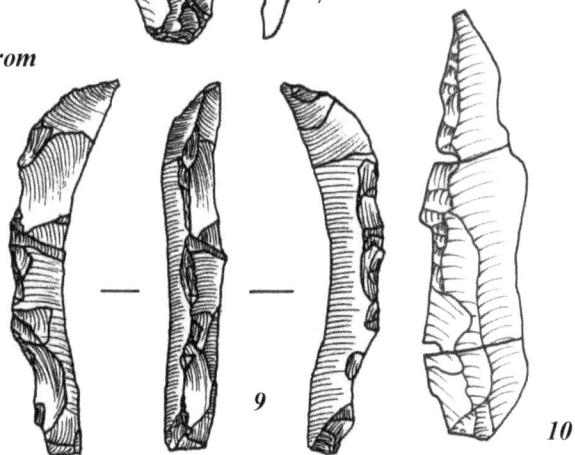

at Zonhoven-Molenheide was mainly organised from two opposed striking platforms (fig. 45, 2-3; 46, 1-2) or less frequently from single platform cores (fig. 45, 1; 46, 3-4). The reduction faces indicate that blade(let)s were mostly, but certainly not always, obtained by a bipolar production. Flakes, in contrast, are more frequently associated with unipolar reduction faces. The analysis of the refits confirms this observation.

Most cores have either natural or plain platforms, or a combination of both. Platforms on the other cores are at least partly facetted. These are generally not very regular. Rather, the platforms were facetted striking off flakes of various sizes and from various directions, often with hinging terminations. Their only apparent purpose was to adjust the flaking angle.

When, for analytical purposes, cores are circumferentially partitioned into 4 equal parts, the extension of the reduction face is mostly limited to one side, less frequently to two sides, and rarely to three or four sides. In general, bipolar reduction faces occupy more of the core circumference than unipolar ones. When the back of a core is not

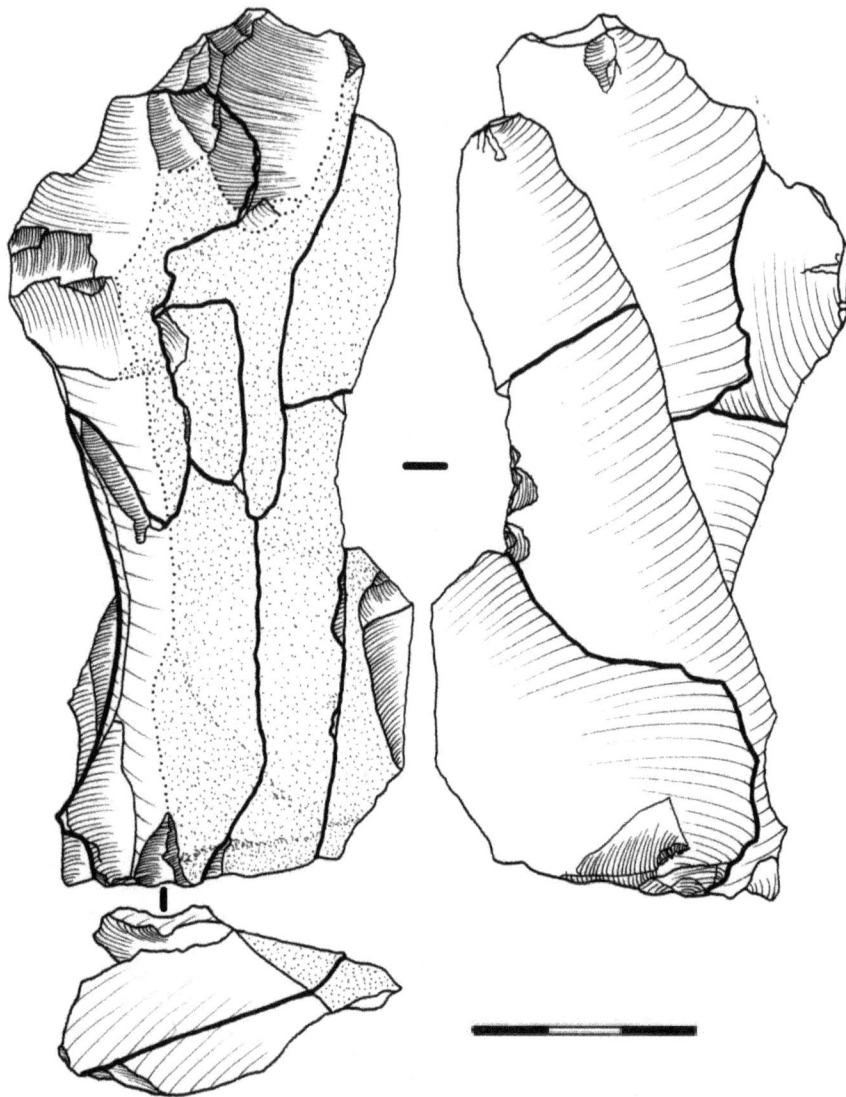

Figure 49 - Refit 125: a series of decortication flakes and blades from an opposed platform core.

consumed by the reduction face, or flaked at an earlier stage, it is mostly left cortical or natural.

Laminar production

As shown by the analysis of the cores and by the choices of the blank types selected for tooling, the flint working at Zonhoven-Molenheide principally targeted laminar production (i.e. blade(let)s). This seems to contradict the apparent lack of pre-planned organisation during the initial shaping of the cores, since blades (*sensu stricto* L>=2W, parallel edges and parallel ribs) are such a strictly defined artefact type that the reduction strategy normally comprises elaborate preparatory stages.

As a consequence of this opportunistic exploitation of the raw material, the initial laminar products at Zonhoven-Molenheide were generally large and thick items, often with a triangular cross-section and thick, unprepared butts (fig. 48).

Laminar production often started with a simple crest preparation and the production of a series of successive cortical blades (fig. 46, 2-3). Long cortical blades are not rare (fig.

45, 3; 46, 2, 7, 12; 48). After the detachment of the opening blade, subsequent specimens were inclined to become gradually shorter. As frosted flint has often been used, frost cracks resulted in the fragmentation of the blade products (fig. 45, 1, 3; 48). Nevertheless, even frosted nodules could produce long blades (fig. 45, 3). Another reason for the rapid cessation of laminar generation, was the quick appearance of hinging accidents, partly as a result of the overall rectilinear profile of the core tables. Platform renewal was one way to deal with these knap-ping accidents.

The succession of the change of knapping direction is highly variable, and was certainly not directed by the knapping accidents alone. From a morphological point of view, the most regular blades were actually produced by a rapid successive extraction along a single core table from two opposed, sometimes rejuvenated, striking platforms. This method could yield a series of up to seven laminar items (fig. 45, 1).

As a consequence of the gradually decreasing dimensions of the core, the terminal stages of the sequences often produced small blades and bladelets. On several occasions, this had to be inferred from the negatives left upon the refitted cores, as rather few bladelets could actually be refitted. No special core treatment seems to have accompanied this bladelet production. Occasionally, bladelets were also sporadically produced (by chance?) during the earlier stages of a reduction sequence.

There is no indication of a serial production of flakes. When flakes were produced their creation is most consistent with laminar production which had for some reason gone wrong at an initial stage (fig. 48).

Previously De Bie (1999) studied some of the characteristics of the debitage products and his remarks can here be confirmed. Blade profiles are straight at Zonhoven-Molenheide, while cortical or irregular butts on blades are extremely rare, and where they occur are on flakes. A high number of linear butts are attested, although plain butts are predominant. Butt thickness is much reduced. Butt dimensions of flakes are on average twice as large as blades. Flaking angles on blades have a mean of 110° whereas on flakes this is close to 100°. In overhang preparation, abrasion is dominant for blades. Bulb striae were seldom observed. Blades from Zonhoven-Molenheide are notable for their small butt, wide flaking angle, abrasion, absence of bulb striae, and presence of tiny contracted percussion ripples. One might conclude from this that softer hammers were used during blade production at Zonhoven. However, some stone hammers are present at the site (fig. 40).

In conclusion, the majority of the reconstructed sequences at Zonhoven-Molenheide reveal a strategy that can be described as a simplified laminar reduction method. Knapping proceeded in a very flexible manner, resulting in an output of mainly unstandardised products.

3.6 – Productivity

Cores are normally discarded when further exploitation is for some reason no longer possible or opportune. Internal flaws or irregularities in the flint material may impede further reduction and knapping may provoke uncontrolled breakage of the volume. Alternatively the reduced core may arrive at a size that can not be further exploited with handheld percussion flaking (fig. 41, 6-7; 42, 3-4; 43, 3). However, the most frequent cause of rejection was the presence of pre-existing frost fractures (fig. 43, 6-7). Overpassing of flakes was also identified as a cause of discard (fig. 41, 2; 44).

There is no doubt that the black flint, type 1, of which only three entirely exhausted cores (fig. 41, 4; 42, 4) were collected is the final knapping output from probably one or a handful of nodules. Unfortunately the refitting of this flint was restricted to smaller pieces.

The occasional absence of cores in comprehensive reduction refits (e.g. Refit 361: fig. 49) probably indicates the intentional export of these items.

It seems that several artefacts of black flint had been knapped elsewhere, transported to the site and discarded.

Although alternative explanations might be applicable, we believe that at least some of the nonreflective large black flint artefacts might represent specimens which had been kept during travel between sites for the occasional production of blanks on the way. Other pieces of black flint were eventually discarded after arrival at a setting where there was sufficient raw material for their replacement.

Selection of blanks for tooling and use

De Bie and Caspar (2000) have stressed that, within the context of a habitation site, it may be postulated that flint knapping primarily served to generate blanks destined for tooling and/or use. It is therefore essential to ask two main questions:

1. What (if any) selection criteria have been adopted? In other words, what blanks were desired for tooling or use, that is what 'end products' were intended to result from the knapping process?

2. Can a variety of preferences be identified among the various tool types, and if so, was this correlated with specific reduction strategies?

Answering the first question is not as easy as it may seem. In fact tool-fabrication by definition implies the modification of a blank. The more a tool is used and sharpened, the less evidence it will retain of the original nature of its blank. Whether a short endscraper was initially made on a flake, a laminar flake or a blade is often hard to define from the tool itself, which is usually found as an intentionally discarded specimen.

In general terms, it seems that few blades were used in tool production. Indeed very few tools from the site attest a blade blank. Once again, tools in black flint constitute the main exception to this. These were made on large flakes or blades. However, bladelets were used more often for the production of atypical Zonhoven points and backed bladelets. Products resulting from the early phases of knapping, or from core maintenance activities, were certainly not excluded from selection for tooling and especially for the production of Zonhoven points.

In this discussion one must of course take into account that many used artefacts do not fall into the category of retouched tools. Unfortunately, while micro wear analysis was attempted for Zonhoven-Molenheide, it did not provide good results because for many artefacts any use wear had been obliterated by the mechanical effects of soil movement (see page 83). It is quite possible that most blades were used as tools.

On the whole, blanks 'consumed' during tooling activities were marked by great formal diversity.

Burnt artefacts

Burnt pieces are numerous on the site (fig. 34). Their distribution is indicated in table 1.

4 - TOOLS

Tool classification sought to employ conventional terminology, largely based on definitions already proposed in the literature (Inizan, Roche, Tixier 1992, Inizan *et al.* 1999; De Bie, Caspar 2000). For the microlithic tools we mainly refer to the approach of G.J. Rozoy (1978). We do not follow Dutch authors (Arts, Deeben 1981) in the use of terms like A-points, B-points, etc., as this may run the risk of confusion since English authors use a similar terminology but with another definition (Barton 1992).

We believe that a simple type list not only provides adequate information, but also is easier to interpret. For these reasons, we have tried wherever possible to avoid creating numerous sub-categories of tool-types. Under a simplified scheme, tools are grouped according to their main morphological characteristics, with any variation being discussed under each category. The tools list is given in table 3.

Some comment is required on our preference for the term "Zonhoven point" . The "Zonhoven Spitze" was defined by G. Schwantes (1928) as a short thin blade, which at its upper end is truncated by a retouch in such a way that the point is situated in the prolongation of the lateral blade edge. H. Schwabedissen (1954: 115) and often also K.J. Narr (1968) used the term also for points which in addition to an oblique proximal truncation also have a basal truncation. According to W. Taute (1968: 182-184) an actual pointed tip is called a B-point. We prefer not to continue the use of W. Taute's term because the term B-point has different meanings in the English and Dutch literature (see above). We prefer to retain the term Zonhoven point ("Zonhoven Spitze") and exclude the irregular trapezes with two truncations. The term "Zonhoven point" is however restricted to points that have the point prepared at the proximal end of the blank. This represented a tightening of the rather broad definition of G. Schwantes (1928), who did not define the place of the point on the blank.

The term "atypical Zonhoven point" will be used to refer to a point which has the general appearance of a Zonhoven point, but where the base is not the distal end of the blank but a fracture. Such pieces could of course be regarded as fragments of an oblique truncated bladelet. However such pieces are quite numerous and their dimensions fit entirely those of typical Zonhoven points.

Some other points with similar dimensions have a distal oblique truncation for which the term "distal Zonhoven point" will be used.

The total excavated collection consists of some 403 retouched tools including 101 Zonhoven points, 42 atypical Zonhoven points, 20 distal Zonhoven points, 7 Ahrensburg points (not all typical), 29 endscrapers, 34 burins, 45 backed bladelets 21 microliths, 44 truncations and a number of retouched pieces., some tanged pieces and tanged points and shouldered points as well as. By contrast, all other types are poorly represented.

4.1 – Endscrapers

This group is composed of 26 endscrapers: 10 are made on flakes (fig. 50, 2, 4), 9 on blades (fig. 50, 1, 3) and 2 on exhausted small cores. None of the scrappers is made of the black flint, type 1. About 10 of the endscrapers (fig. 50, 5-7) are small, some of them being thumbnail scrapers. Many endscrapers have some cortex on their dorsal surface (fig. 50, 1, 4, 11). Some of the endscrapers are made on a unidirectional blank while others on a bidirectional blank.

The morphological features of the scraper edges appear to be fairly uniform. The edge is almost always semi-circular in plan and placed at the distal end of the blank.
A single sidescraper is present.

4.2 – Borer

There is only a single piece identified as borer.

4.3 – Burin

Burins are well represented. They are dihedral burins or burin on a truncation, most of them with an complex history of utilisation and sharpening, often resulting in double or multiple burins.

Most burins were collected in the NW-sector, where several of the larger burins, that have often been sharpened, are made of the black flint, type 1. Their blank is often a cortical flake. A number of those burins were refitted. They generally started as large tools, being reduced during numerous reuse stages. Most burins were rejected as dihedral burins but they often have passed trough a stage as burin on a truncation. Sharpening, with the production of burin spalls was often applied.

A double dihedral burin (fig. 50, 13) from the NW-sector has three burin spalls, all from the same NW-sector. The burin was made on a cortical flake and the burins are edge burins.

The distal, cortex bearing, end of a rather thick flake in black flint, type 1, was retouched, creating an endscraper or a convex truncation, of which two retouching chips were refitted (fig. 50, 8). A burin blow on the truncation was plunging, sectioning the flake, creating a burin on a convex truncation (fig. 50/11). At the distal end of the burin facet another burin blow was applied creating a dihedral burin (fig. 50, 8). The large burin spall (fig. 50, 10) was thinned by some retouches on the dorsal surface and a new burin blow was give at the end of the earlier burin facet, creating a dihedral burin. The other end of the flake was also thinned by some dorsal rather flat retouches (fig. 50, 8). It was removed by a burin blow the facet of which was used at least twice for the production of an dihedral burin (fig. 50, 10). All elements of the refit sequence were collected in the NW-sector.

From a distal blade fragment in black flint type 1, two distal transversal burin spalls were detached (fig. 51, 7). A first edge dihedral burin was created by a burin blow on the dorsal side of the blade. At the proximal part of the blade a slightly concave truncation was created and a burin blow applied, which was plunging, creating a small edge burin on a truncation and disabled the distal burin. At the distal part of the blade a new dihedral burin was created. At the right hand side of the truncation another burin blow was applied which created a "burin plan". All elements of the refit sequence were collected in the NW-sector.

A angle dihedral burin (fig. 51, 1) made from a flake in black flint type 1, was sharpened many (at least 7) times, leaving a series of burin spalls which could be refitted. The

Table 3 - List of tools according to their presence in one of the sectors.

	NW	NW%	NE	C	P	P%	R	SE	SW	Total	Total %	Total	Total %
Endscraper on a blade				2	3	2.44	1	1	2	9	2.34		
Endscraper on a flake	2	1.50		1	1	0.81	1	1	4	10	2.60		
Thumbnailscraper				2					1	3	0.78		
Denticulated endscraper					1	0.81				1	0.26		
Core endscraper	1	0.75		1						2	0.52		
Sidescraper								1		1	0.26		
Scrapers												26	6.77
Borer								1		1	0.26		
Borers												1	0.26
Burin dihedral burin	6	4.51					1	1		8	2.08		
Burin on a truncation	11	8.27	2		3	2.44		1		17	4.43		
Burin on a break	2	1.50				0.00	1			3	0.78		
Burin on a lateral retouch	1	0.75				0.00				1	0.26		
Burin plan	3	2.26				0.00				3	0.78		
Multiple burin on a truncation		0.00			1	0.81				1	0.26		
Double burin on a break	2	1.50								2	0.52		
Dihedral burin - endscraper	1	0.75								1	0.26		
Burins												28	7.29
Pointed straight backed bladelet					1	0.81	1	1		3	0.78		
Pointed convex backed bladelet	4	3.01		2			1			7	1.82		
Pointed backed bladelet	3	2.26			4	3.25				7	1.82		
Backed points												17	4.43
Straight backed blade(let)				1	2	1.63	1			4	1.04		
Convex backed bladelet	1	0.75			1	0.81		1		3	0.78		
Backed bladelet fragment	1	0.75		3	2	1.63		2	8	16	4.17		
Truncated backed bladelet				1						1	0.26		
Partially backed bladelet	1	0.75			3	2.44				4	1.04		
Backed flake	2	1.50		3	1	0.81		1		7	1.82		
Backed element												35	9.11
Zonhoven point	33	24.81	1		53	43.09	6	11	5	109	28.39		
Atypical Zonhoven point	6	4.51	1		11	8.94	2	7	4	31	8.07		
Distal Zonhoven point	4	3.01	1	1	4	3.25	2	5		17	4.43		
Ahrensburg point	4	3.01			1	0.81				5	1.30		
Atypical Ahrensburg point	1	0.75			1	0.81				2	0.52		
Shouldered point	1	0.75								1	0.26		
Points												165	42.97
Trapeze	4	3.01		1	3	2.44	2			10	2.60		
Triangle	3	2.25	2		6	4.88		1		12	3.12		
Segment	1	0.75							1	2	0.52		
Geometric microliths												24	6.25
Truncated blade(let)	11	8.27	1		10	8.13		6		28	7.29		
Truncated flake	3	2.26			1	0.81		5		9	2.34		
Double truncated blade(let)	2	1.50								2	0.52		
Truncations												39	10.16
Notched blade	3	2.26			1	0.81				4	1.04		
Notched flake					1	0.81				1	0.26		
Blade(let) broken in notch	2	1.50			1	0.81			1	4	1.04		
Denticulated flake					1	0.81				1	0.26		
Splittered blade							1			1	0.26		
Retouched blade(let)	8	6.02		1	4	3.25	1	4		18	4.69		
Retouched flake	6	4.51		1	2	1.63		2	1	12	3.13		
Divers												41	10.68
Total	133	100.00	6	22	123	100.00	20	52	28	384	100.00	384	100.00

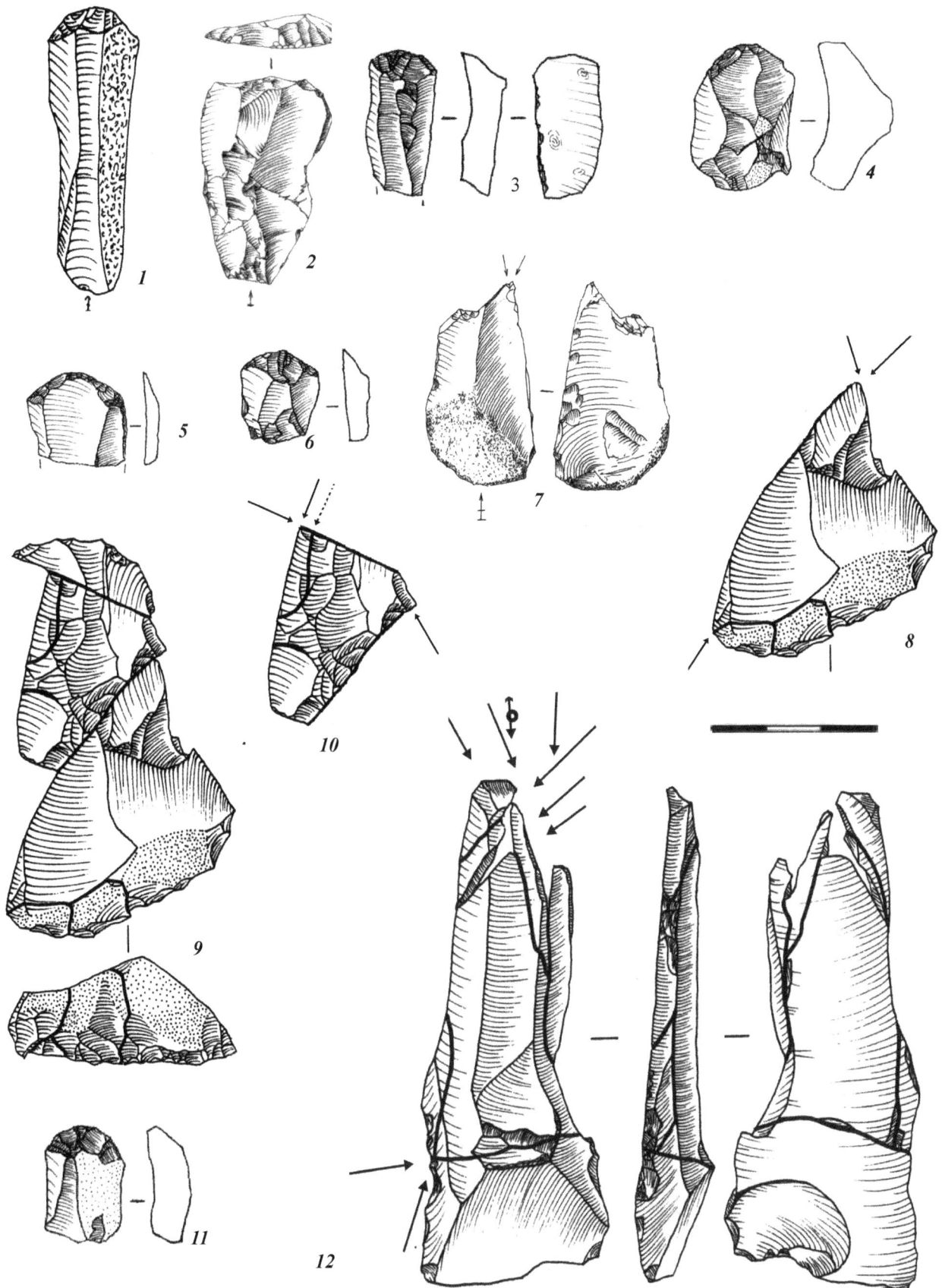

Figure 50 - 1-3: endscraper on a blade; 4: endscraper on a flake; 5-6, 11: thumbnailscraper; 7: dihedral burin; 8: double dihedral burin-endscraper; 9: refitted artefact with burins 8 and 10; 10: double dihedral burin; 12: refitted double dihedral burin. (1: from P; 2, 8-11: from NW; 3-4, 7: from SW; 5-6 from C; 12 from SE).

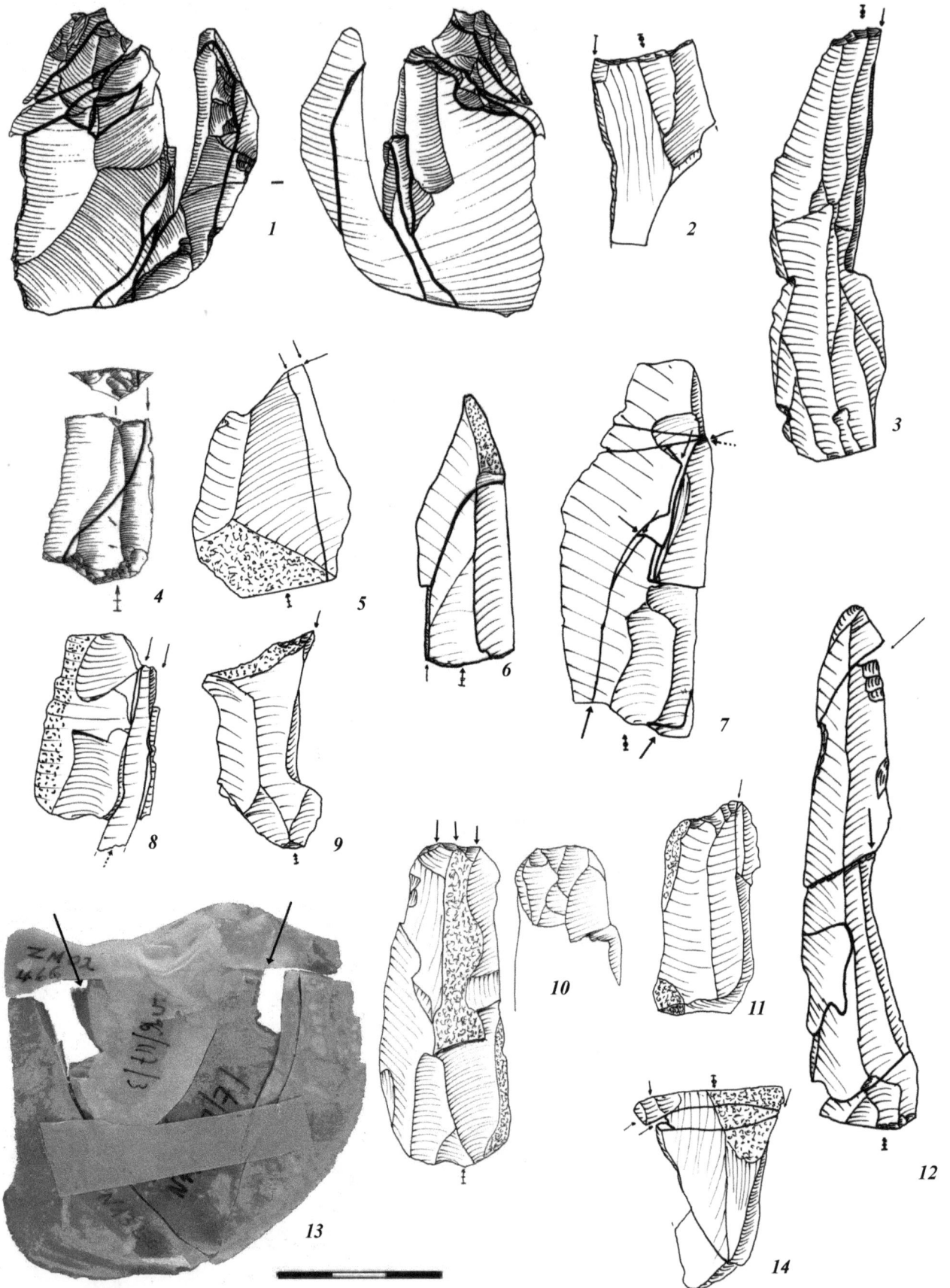

Figure 51 - Burins 1: dihedral burin; 2, 4, 9, 11: angle burin on a concave truncation, 3: angle burin on a straight truncation; 5: dihedral burin with refitted burin spall. 6: burin on a break; 7: refitting of successive burins; 8: burin on a splittered blade; 10: burin plan; 12: refitting of a truncated blade and a burin; 13: ventral view of a double edge dihedral burin; 14: double dihedral burin; (1,4-8, 10-12, 14 from NW; 2-3 from P; 13 from R; 9 from SW).

burin and its spalls were deposited in the NW-sector. A burin spall was refitted to a "déjeté" dihedral burin (fig. 51, 5).

A double angle dihedral burin (fig. 51, 14) was made on a flake blank in flint type 2. Several burin spalls were refitted. The burin and spalls come from the NW-sector.

A double dihedral angle burin on a cortical flake blank in grey speckled flint, type 4, has been frequently sharpened; several burin spalls were refitted (fig. 51, 13).

A blade in black flint type 1, was transformed into a angle burin on a concave truncation (fig. 51, 4). The refitted spall is plunging.

An angle burin on a concave truncation (fig. 51, 2) in black flint, type 1, was recovered from the P-sector.

A plunging burin blow on a cortical flake blank in black flint type 1, created a burin on a concave truncation (fig. 51,11), was.

An angle burin on a truncation (fig. 51, 3) was obtained from a long blade in black flint type 1. The burin was collected in the P-sector.

Another burin (fig. 50, 7) on a cortical flake in black flint type 1, was first an edge burin on a concave truncation but in its present state has been sharpened as a dihedral burin.

A cortical distal blade fragment in the black flint type 1, served as a blank for a burin on a break (fig. 51, 6). The burin refits to a large cortical truncated flake (Refit 257ter, fig. 55, 10).

A long blade in black flint type 1 has been fragmented into four parts, all recovered from the NW-sector. The distal part has received a burin blow which did not create a strong burin. After a rather flat retouch on the distal right edge a new burin blow created a good burin tip (fig. 51, 12). The medial part of the blade obtained an inverse straight truncation, which at the right edge served for the creation of a edge burin on a truncation.

The distal cortex of a flake was slightly reworked by a truncation and a burin blow, creating an angle burin on a truncation (fig. 51, 9).

A cortical flake in black flint type 1from the NW--sector is now a splintered piece (fig. 51, 8 for a view of its ventral face), but several burin spalls from a earlier situation as "burin plan" are refitting.

Backed blade(let)s

A significant number of backed bladelets are present. They were collected from all across the excavated area, although relatively few came from the NW-sector. Most are small to very small fragments. Backing is most often obtained by an abrupt retouch.

Several types are present but there is no systematic approach to this type of tools. Nearly all have bladelets and not blades as blanks. Most items are fragments. There is a single blade blank (fig. 52, 12), collected from the C-sector. It might possibly be a fragment of a Tjonger point.

Among the backed blade(let)s about one third are pointed. Some items are pointed straight backed points (e.g. fig. 52, 1, here with a fine retouch on the left edge). Most

are pointed convex backed microlithic points, the pointed tip often being proximal (fig; 52,2-5). Three are very fine (fig. 52, 14, 17-18), two of them being produced in Wommersom quartzite (fig. 52, 14, 18).

Most items are fragments (fig. 52, 9, 13) of backed bladelets where sometimes backing is only partial (fig. 52, 6-8, 10, 16). Some are straight backed bladelets (fig. 52, 15). Others have a rather convex backing. Occasionally the proximal end obtained an oblique truncation (fig. 52, 11).

Two artefacts could eventually be interpreted as Krukowski microburins.

Zonhoven points

Small points are the most characteristic tools in the assemblages. As already discussed, a distinction was made between Zonhoven points, atypical Zonhoven points and distal Zonhoven points. All of them have been obtained through an oblique truncation of a flake or occasionally of a bladelet.

Zonhoven point (fig. 52, 33-35; 53; 54, 1-9), with 109 items, are the most numerous tool type on the site. All of them are characterised by a proximal truncation while the distal end of the blank forms the basis of the point. The morphology of the point is clearly not standardised as apparently very different blanks were used to produce the points. Still it seems that length and width of the points are more or less standardised with a length generally between 2 and 3 cm. The oblique truncation, which is mostly slightly concave, can be applied on the left blank edge (fig. 52, 33-35; 53, 1-28) or on the right (fig. 53, 29-35; 54, 1-9). The lateralisation is therefore clearly not constant, although there seems to be a slight preference for the left lateralisation.

Atypical Zonhoven points (fig. 54: 10-21) are, with 31 items, also an important tool type. They are most often made on a bladelet, sometimes on a flake, by a flex-ion break opposite the truncation. It is presumed that the fracture was intentional to obtain the normal dimension of a Zonhoven point. This procedure is visible in a refit of such a point with the distal part of the blank (fig. 54, 15). The dimensions of atypical Zonhoven points are similar to those of the Zonhoven points. The point base is normally formed by an unretouched break.

Some points, here specified as distal Zonhoven points, have been obtained by applying a distal truncation near the proximal end of the blank. As in the other Zonhoven points, the tip is formed by an oblique truncation. In such cases the blank butt forms the base of the point (fig. 54, 22-34). Less frequent than the other Zonhoven points, they still are quite well represented.

The final morphology of the atypical Zonhoven points and the distal Zonhoven points is quite similar to that of "real" Zonhoven points. They were presumably used in a similar way.

Ahrensburg points

Ahrensburg points (fig. 55, 1-3) are not numerous and not very characteristic. As with the Zonhoven points, their tip is always situated at the proximal end of the blank. The central tang is short and small. Some points have been identified as atypical Ahrensburg point (fig. 53, 19).

However, some of the trapezes (fig. 52, 26-27, 30-32) have

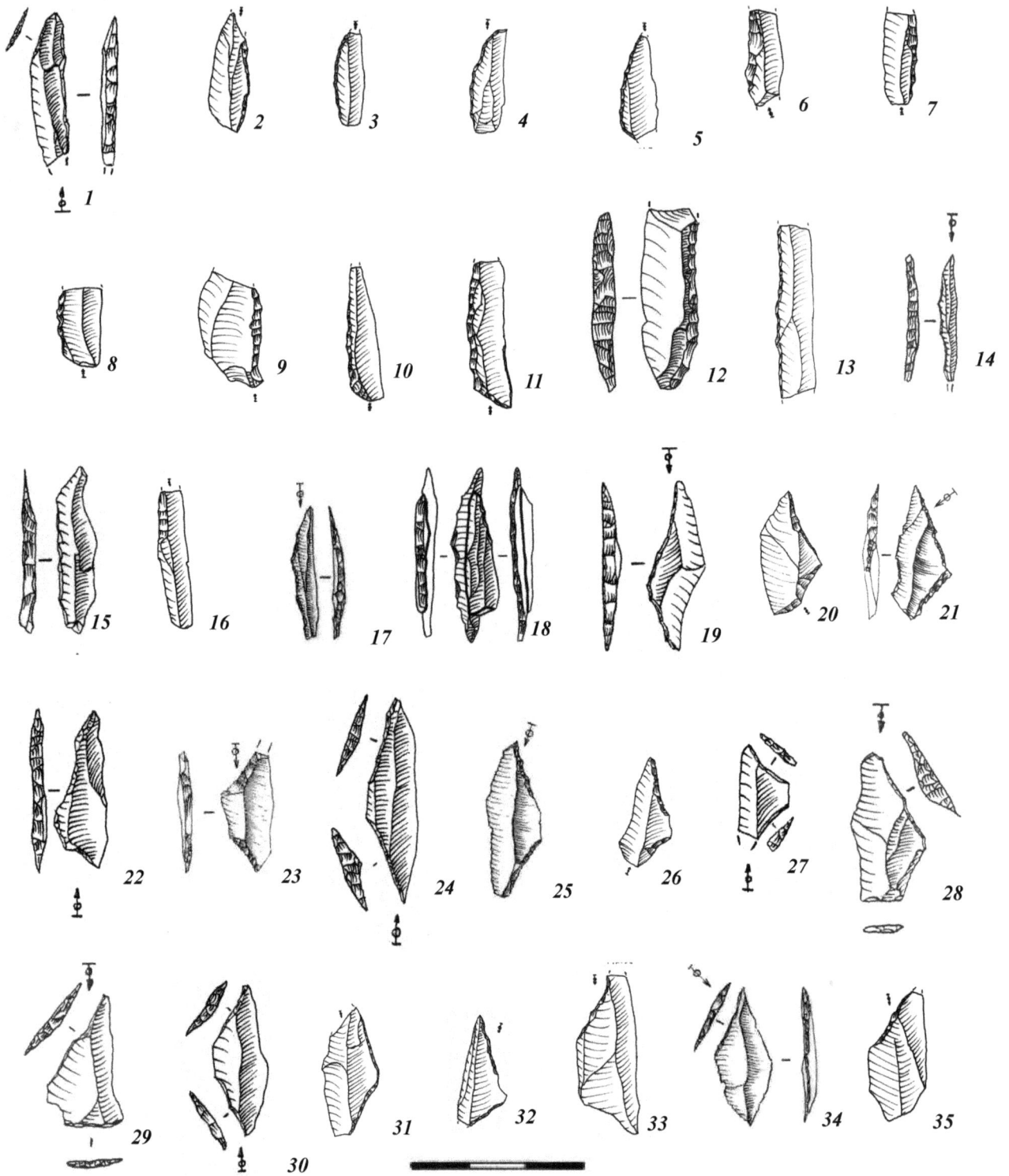

Figure 52 - 1-11, 13-17: backed bladelets; 12: backed blade fragment; 18 segment refitted to a backed bladelet; 19-22: triangle; 23-32: trapeze; 33-35: Zonhoven point. (1, 26: from R; 2, 8, 13, 18, 30: from SW; 3, 15-16, 22, 27, 31-32, 35: from P; 4, 19, 33: from SE; 5 ,7, 12, 14, 17, 20-21, 23-25, 28-29, 34: from NW; 6: from NE; 7, 9-12: from C).

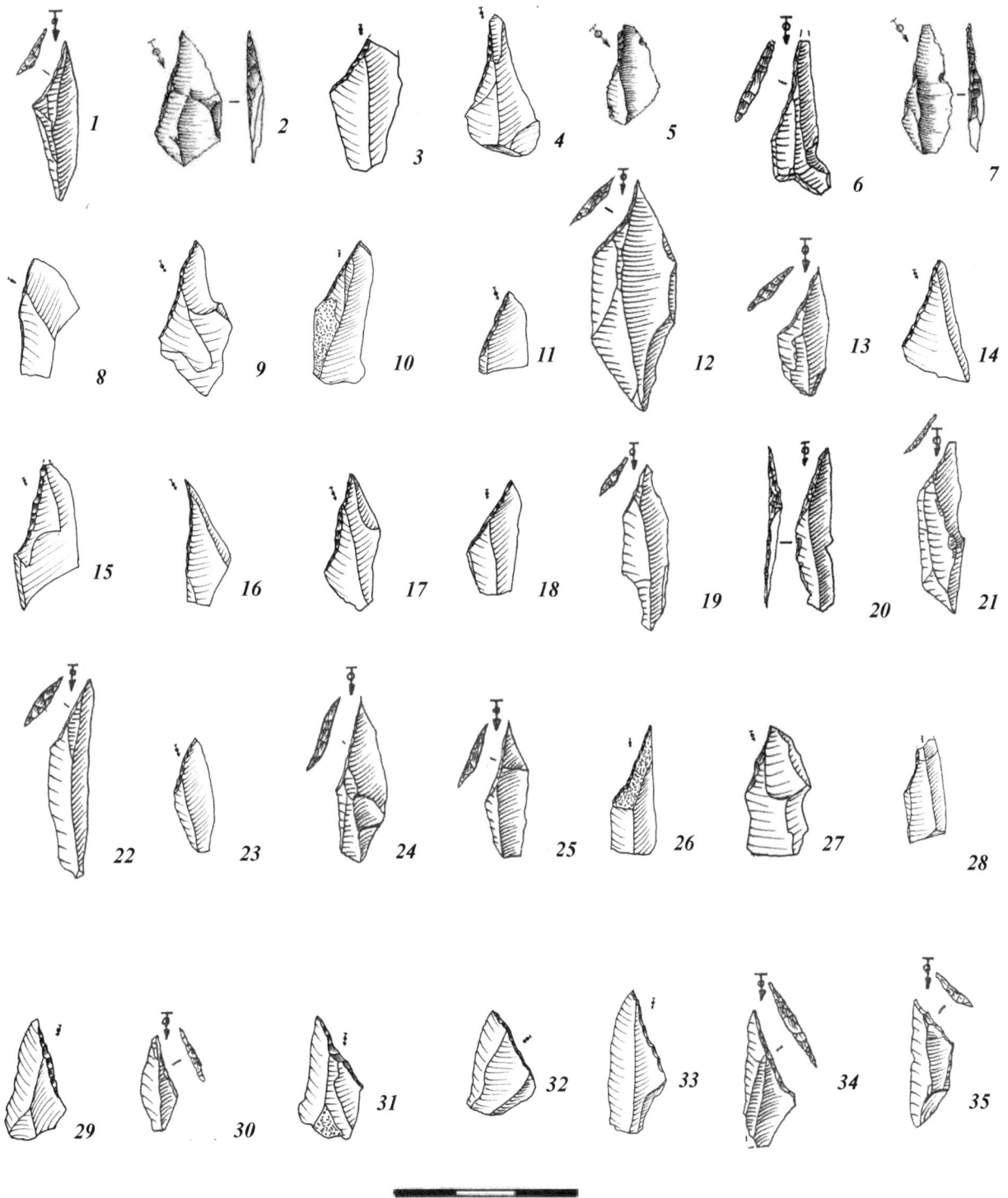

Figure 53 - 1-18, 20-35: Zonhoven points; 19: atypical Ahrensburg point. (1, 6, 11-13, 15-23; 28-35: from P; 2-5, 7-10, 14, 26-27: from NW; 24 from SE; 25: from R).

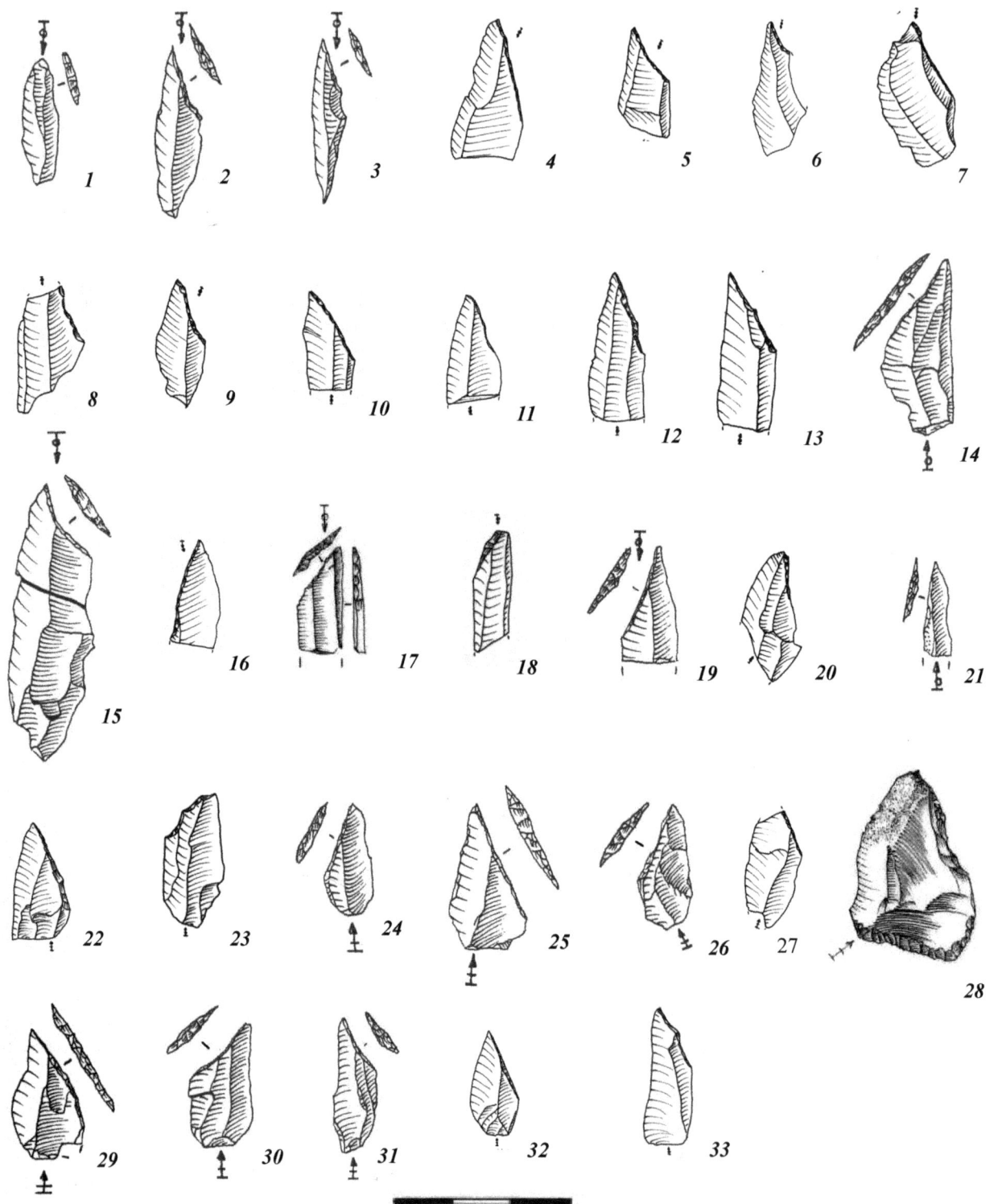

Figure 54 - 1-9: Zonhoven point; 10-21: atypical Zonhoven point; 22-33: distal Zonhoven point. (1, 4, 21, 28: from R; 2: from SW; 3, 5-8, 13-14, 17, 22-23, 29, 32: from NW; 9, 12, 16, 18-19, 24, 30-31, 34: from P; 10-11, 15, 20, 25-27, 33: from SE; 21: from R).

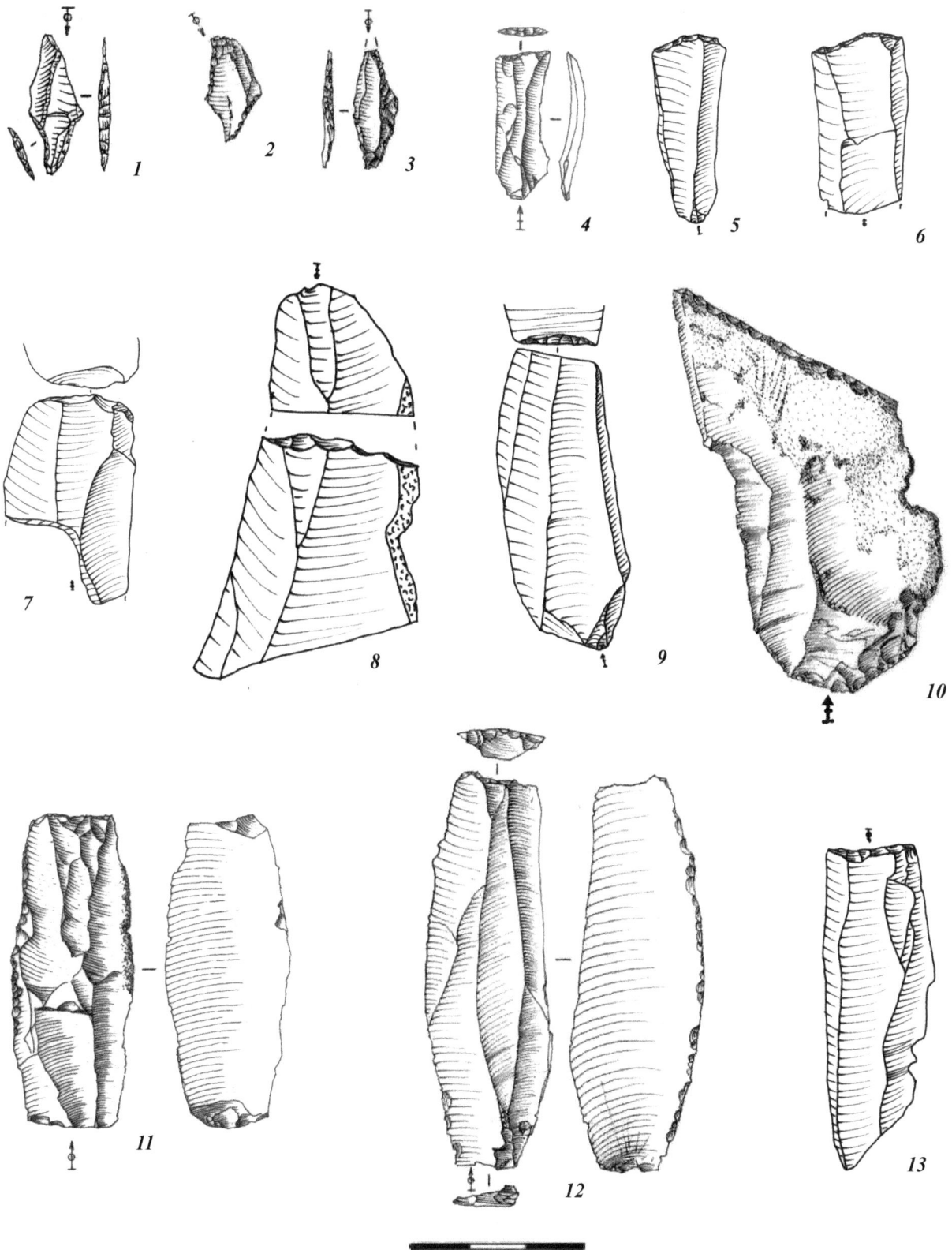

Figure 55 - 1-3: Ahrensburg point; 4-13: truncated pieces. (1-4, 8-12: from NW; 5: from SE; 6, 13: from P; 7: from R).

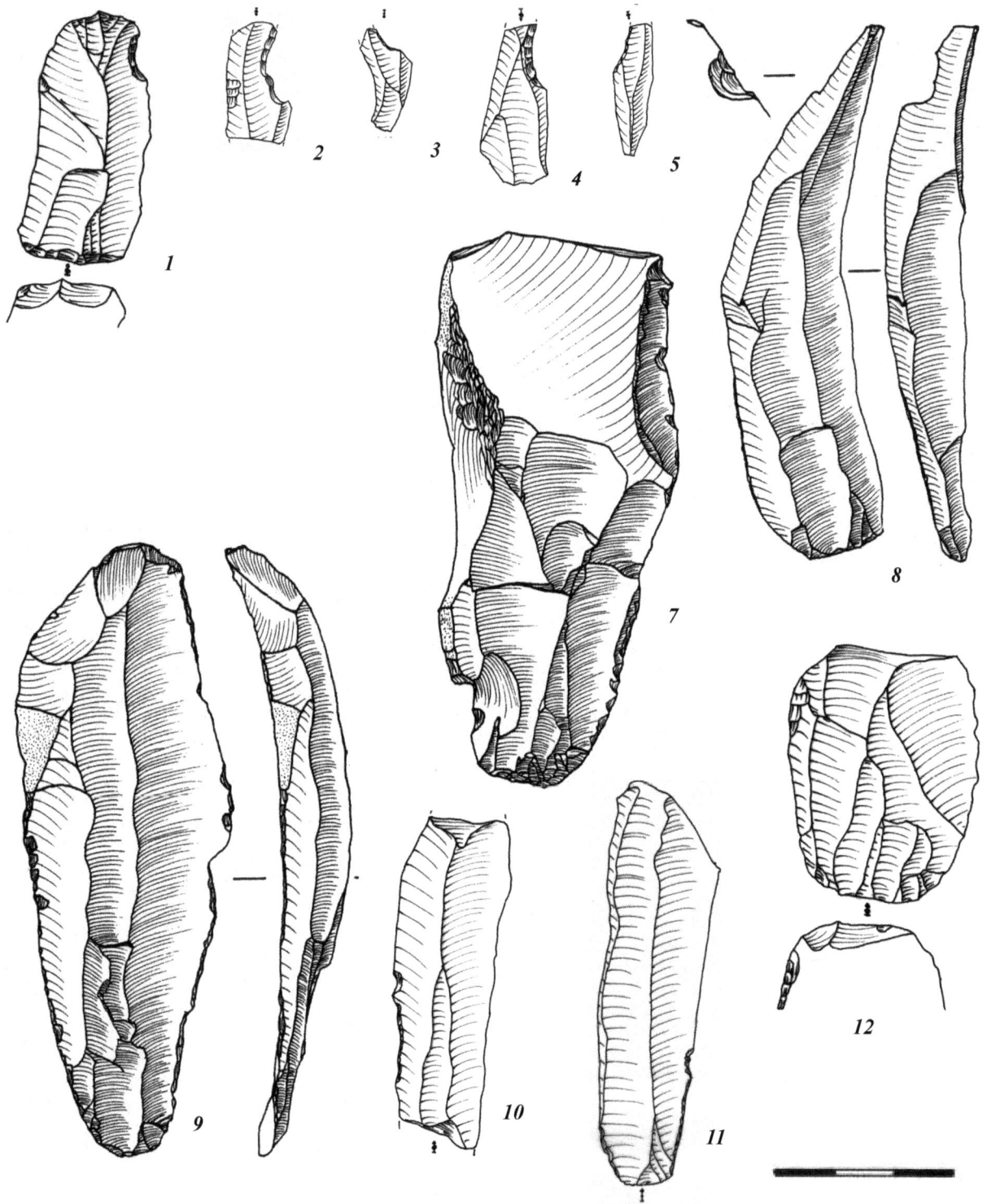

Figure 56 - 1: splittered retouched blade; 2-5: bladelets and flake broken in a notch; 6: retouched core rejuvenation; 7: retouched flake; 8-11: retouched blade; 12: splittered disc. (1-2, 6-7, 10-12: from NW; 3-5, 8: from P; 9: from SE).

a morphology very similar to that of an Ahrensburg point. However, in such trapezes the tang is not fully individualised. As with a Zonhoven point, an Ahrensburg point tip is obtained on the proximal part of the blank.

Microliths

Microliths are present, but are normally slightly different from typical Mesolithic types in their symmetry.

There is a single item in Wommersom quartzite (fig. 52, 18), which refits to the backed bladelet noted above. Triangles (fig. 52, 19-22) are present but are not symmetrical and were not necessarily made on bladelets. Trapezes (fig. 52, 23-32) are more numerous, but again most are irregular trapezes, aside for a few that resemble late Mesolithic items (fig. 52, 24-25). For none of the microliths has the presence of a "picquant trièdre" been observed, an observation which is confirmed by the absence of microburins.

Truncations

Truncated pieces (fig. 46, 2; 55, 4-13) are well represented and mainly produced on a blade blank. The large elongated pieces are often made in black flint type 1. A large cortical flake in black flint has a very regular straight truncation (fig. 55, 10). This piece was refitted with a burin (fig. 51, 5; fig. 65). Occasionally an inverse retouch was used to thin the truncation (fig. 55, 7 and 11), but sometimes the truncation is only inverse (fig. 55, 9). Some truncations have apparently been prepared on a blank that had previously been sectioned by a flexure break (fig. 55, 8). The truncation is most often somewhat concave (fig. 55, 4, 12-13) but can sometimes be convex (fig. 55, 5-6) or straight (fig. 55, 8). An inverse irregular retouch is present on a truncated blade (fig. 55, 12). A double truncated blade (fig. 55, 11) also has a discontinuous left edge retouch.. An inverse irregular retouch is present on a truncated blade (fig. 55, 12). A double truncated blade (fig. 55, 11) has also a discontinuous left edge retouch.

Miscellaneous

Notched pieces, sometimes with inverse notch and denticulates are rare (fig. 56, 8). However, this observation might be conditioned by the fact that not all artefacts were in a very fresh state of preservation. On such pieces notches have been disregarded. A long blade with a ventral notch (fig. 56, 6) is made of Wommersom quartzite.

Some blades and bladelets have been broken at a notch (fig. 65, 2-5), but none of them displays a "piquant trièdre".

Splittered pieces (fig. 56, 1) are present but not numerous. One of them is a rather thin oval flake, nearly a discus, of which one end has been splittered (fig. 56, 12).

Retouched blades (fig. 56, 9-11) and flakes (fig. 56, 7) frequently show traces of intensive use. The large pieces (fig. 56, 7, 9) are made from the black flint type 1.

Two refitted blades have been obtained from a frosted fragment.

The lower one (fig. 56, 6) resembles a Tjonger point, but is not one. The blade edge has been regularised by an inverse, steep, irregular and somewhat denticulated retouch, which should be interpreted as a core edge preparation. The base is rounded.

61

5 - REFITTING

At Zonhoven-Molenheide refitting resulted in 532 refitted groups and a total of 1,825 artefacts. Refitting was performed by the late Chris Peleman. Her premature death unfortunately prevented the completion of the refitting procedure.

In this study the method of presentation applied at Rekem (De Bie, Caspar 2000) served as an example. However, only refits with at least 5 artefacts are discussed in the following descriptive analysis. In most cases the core abandonment state (item 4) is unknown, but all cases where it could be discerned are mentioned. Informative drawings of some other refits are included.

Refit 11 (fig. 45, 1)

Refit state: A conjoinment of 15 blanks, illustrating part of a reduction sequence.

Original nodule: Flint type 4. Dimension unknown (min. length 8 cm).

Reduction sequence: All the artefacts of this sequence were detached from a single platform core, all from the same platform.

Productivity: A range of blades. An irregularity inside the nodule and probably a frost crack at the same place resulted in the termination orthe breakage of some of the blades.

Quality of knapping: Mainly excellent blade production.

Spatial layout: All artefacts were collected in the P-sector.

Refit 12 (fig 46, 4)

Refit state: A conjoinment of 5 blanks.

Original nodule: Flint type 4. Dimension unknown (min. length 8 cm).

Reduction sequence: All the artefacts in this sequence were detached from a single platform core with a renewal phase for the production of blades, All are cortical.

Productivity: A range of blades.

Quality of knapping: excellent unidirectional blade production.

Spatial layout: The artefacts originate mainly from the P-sector but also from the NW and NE-sectors.

Refit 13

Refit state: A series of 5 blanks.

Original nodule: Flint type 4. Dimension unknown (min. length 8 cm). The cortex is slightly rolled but still quite fresh. It is probably the same nodule as refit 12, but here blades attest a bidirectional flaking.

Reduction sequence: All the artefacts of this sequence were detached from an opposed platform core with at least one renewal phase for the production of blades, all cortical.

Productivity: A range of blades.

Quality of knapping: good unidirectional blade production.

Spatial layout: All artefacts originate from the P-sector.

Refit 38

Refit state: A conjoinment of 5 blanks, illustrating part of a reduction sequence.

Original nodule: Flint type 4. Dimension unknown (min. length 5 cm).

Reduction sequence: All the artefacts of this sequence were detached from a single platform core with a renewal phase of the platform for the production of blades, first one being cortical.

Quality of knapping: A range of broken blades and flakes.

Spatial layout: A single flake from sector NE; the others from P-sector.

Refit 47

Refit state: A conjoinment of 8 blanks, illustrating part of a reduction sequence.

Original nodule: Flint type 4. Dimension unknown (min. length 8 cm).

Reduction sequence: Partially from the outside of the nodule with still lot of cortex. All the artefacts of this sequence were detached from a single platform core. As the nodule apparently had many frost cracks, all flakes were broken.

Productivity: A range of flakes and chunks. There is a multiple burin on a truncation with a burin spall, quite anterior to the present burin shape.

Quality of knapping: Mainly chunk production.

Spatial layout: All artefacts originate from the P-sector.

Refit 49

Refit state: A conjoinment of 7 blanks illustrating part of a reduction sequence.

Original nodule: Flint type 3. Dim. unknown (min. length 8 cm).

Reduction sequence: This set represents a sequence of bidirectional laminar reduction from a opposed platform core. There are at least two platform renewals. All were detached during the initial shaping of the nodule and still present some cortex.

Core/abandonment: Unknown.

Productivity: A series of blades, some of which are fragmented.

Quality of knapping: Successful laminar production, slightly hampered by frost breaks.

Spatial layout: All elements were situated in the P-sector except a long blade from the NW-sector.

Refit 51

Refit state: A conjoinment of 6 blanks, illustrating part of a reduction sequence.

Original nodule: Flint type 3. Dim. unknown (min. length 6 cm).

Reduction sequence: This set represents a sequence of unidirectional laminar reduction. There are three platform renewals and a unilateral crest.

Productivity: A series of laminar flakes, all with some cortex.

Quality of knapping: Successful laminar Production. Several butts were intentionally prepared (facetted).

Spatial layout: A proximal blade fragment originates from the NE-sector. The others are from the P-sector.

Refit 55

Refit state: A conjoinment of 13 blanks, illustrating part of a reduction sequence.

Original nodule: Flint type 3. Dim. unknown (min. length 6 cm).

Reduction sequence: This set represents a sequence of unidirectional laminar reduction. There are at least two platform renewals. All were detached during initial shaping of the nodule.

Productivity: A range of cortical flakes, a bilateral crested flake, and short blades.

Quality of knapping: Successful laminar production hampered by frost cracks.

Spatial layout: All elements were collected in the P-sector.

Refit 56-57

Refit state: A conjoinment of 7 blanks, illustrating the problems created by the use of frosted flint.

Original nodule: Flint type 3. Dim. unknown (min. length 6 cm).

Reduction sequence: This set illustrates an attempt to knap an entirely frosted flint nodule. Two opposed platforms were created in an attempt to produce a cortical blade after some unilateral cresting. The attempt failed because the blade was plunging due to the presence of a frost crack. Another attempt sought to reform the remaining chunk into an opposed platform core. This chunk scattered into several pieces.

Productivity: flake, blade and chunk fragments
.
Quality of knapping: Unsuccessful because of frosted flint.

Spatial layout: All elements were collected in the P-sector.

Refit 58

Refit state: A conjoinment of 5 blanks, illustrating part of a reduction sequence.

Original nodule: Flint type 3. Dim. unknown (min. length 5 cm).

Reduction sequence: This set represents a sequence of unidirectional laminar reduction. There are at least three platform renewals. All were detached during the initial shaping of the nodule.

Productivity: A range of elongated flakes.

Quality of knapping: Unsuccessful laminar production, hampered by frost cracks.

Spatial layout: All elements were collected in the P-sector.

Refit 6l

Refit state: A conjoinment of 6 blanks.

Original nodule: Flint type 3. Dim. unknown (min. length 6 cm).

Reduction sequence: This set represents a sequence of unidirectional laminar reduction. There are at least two platform renewals. All were detached during the initial shaping of the nodule.

Productivity: cortical blade and blade fragments (probably because of the use of frosted flint).

Quality of knapping: Successful blade and laminar flake production.

Spatial layout: All elements were collected in the P-sector.

Refit 64

Refit State: A conjoinment of 9 blanks.

Original nodule: Flint type 3. Dim. unknown (min. length 6 cm).

Reduction sequence: This refit represents a sequence of unidirectional laminar reduction. All were detached during the initial shaping of the nodule.

Productivity: cortical blade and blade of flake fragments. A single burin on a truncation with three refitted burin spalls and a multiple burin on a truncation are included.

Quality of knapping: Successful blade and laminar flake production.

Spatial layout: The elements were collected in the the P-sector, except the multiple burin on a truncation, which came from the C-sector.

Refit 67

Refit state: A conjoinment of 10 blanks, illustrating the decortication of a cobble with problems created by the use of frosted flint.

Original nodule: Flint type 3. Fresh pitted cortex of an frosted flint nodule. Dim. unknown (min. length 9 cm).

Reduction sequence: This set illustrates an attempt to knap an entirely frosted flint nodule. The nodule was prepared by detaching a sequence of centripetal flakes, thus clearing the cortex. These flakes were not refitted. From a single platform several cortical blades were removed but all broke up into several parts along frost cracks..

Productivity: chunks, chips, flakes and blade fragments

Quality of knapping: Unsuccessful because of frosted flint.

Spatial layout: All elements were collected in the P-sector.

Refit 71

Refit state: A conjoinment of 5 blanks, illustrating part of a reduction sequence.

Original nodule: Flint type 4. Dim. unknown (min. length 4 cm).

Reduction sequence: This set represents a sequence of uni-directional laminar reduction from a single platform core.

Productivity: A series of elongated flakes.

Quality of knapping: Successful laminar production.

Spatial layout: All elements were collected in the P-sector.

Refit 74

Refit state: A conjoinment of 13 blanks.

Original nodule: Flint type 4, frosted fragment. Dim. unknown (min. length 6 cm).

Reduction sequence: The artefacts of this sequence were detached from a single platform.

Core/abandonment: An opposed platform core.

Productivity: A range of cortical flakes, flakes, elongated flakes and blades fragmented because of frost cracks.

Quality of knapping: Mainly unidirectional flake production.

Spatial layout: All elements were collected in the P-sector.

Refit 76 (fig. 44, 3)

Refit state: A conjoinment of 7 blanks.

Original nodule: Flint type 1. Dim. unknown (min. length 16 cm).

Reduction sequence: This set represents a sequence of

Figure 57 - Distribution of artefacts belonging to Refit 76.

bidirectional laminar reduction with a unilateral crested cortical blade and several rejuvenations of both platforms. One platform has been refreshed at least once. It is probably one of the initial stages in the reduction of a (the?) large black flint fragment type 1, which has a rolled cortex and some frosted surfaces. Many of the artefacts in black flint could have belonged to this nodule.

Productivity: Two long cortical blade and a series of shorter blades of which the last one is slightly retouched. Some blades are broken, but apparently not because of frost cracks.

Quality of knapping: Successful blade and short blade production.

Spatial layout: fig. 57.

Refit 90

Refit state: A conjoinment of 5 blanks.

Original nodule: Flint type 2. Dim. unknown (min. length 3 cm).

Reduction sequence: A series of 5 elongated flakes.

Productivity: A range of flakes from a single platform core.

Quality of knapping: Unidirectional knapping.

Spatial layout: All elements were collected in the P-sector.

Refit 115

Refit state: A conjoinment of 8 blanks.

Original nodule: Flint type 6 with thick rough white cortex, clearly not rolled. Dim. unknown (min. length 5 cm).

Reduction sequence: This set represents a sequence of unidirectional laminar reduction on its core. Refits from the other platform have not been found. There are at least two platform renewals. All were detached during the initial shaping of the nodule.

Core/abandonment: An opposed platform core.

Productivity: a plunging flake and laminar flakes).

Quality of knapping: Poor laminar flake production.

Spatial layout: fig. 58.

Refit 117

Refit state: A conjoinment of 6 blanks.

Original nodule: Flint type 6 with thick rough white cortex, clearly not rolled, but frost cracked. Dim. unknown (min. length 6 cm.

Reduction sequence: Four successive laminar flakes that broke up because of the presence of a frost crack.

Productivity: laminar flakes.

Quality of knapping: Poor laminar flake production.

Figure 58 - Distribution of artefacts belonging to Refit 115.

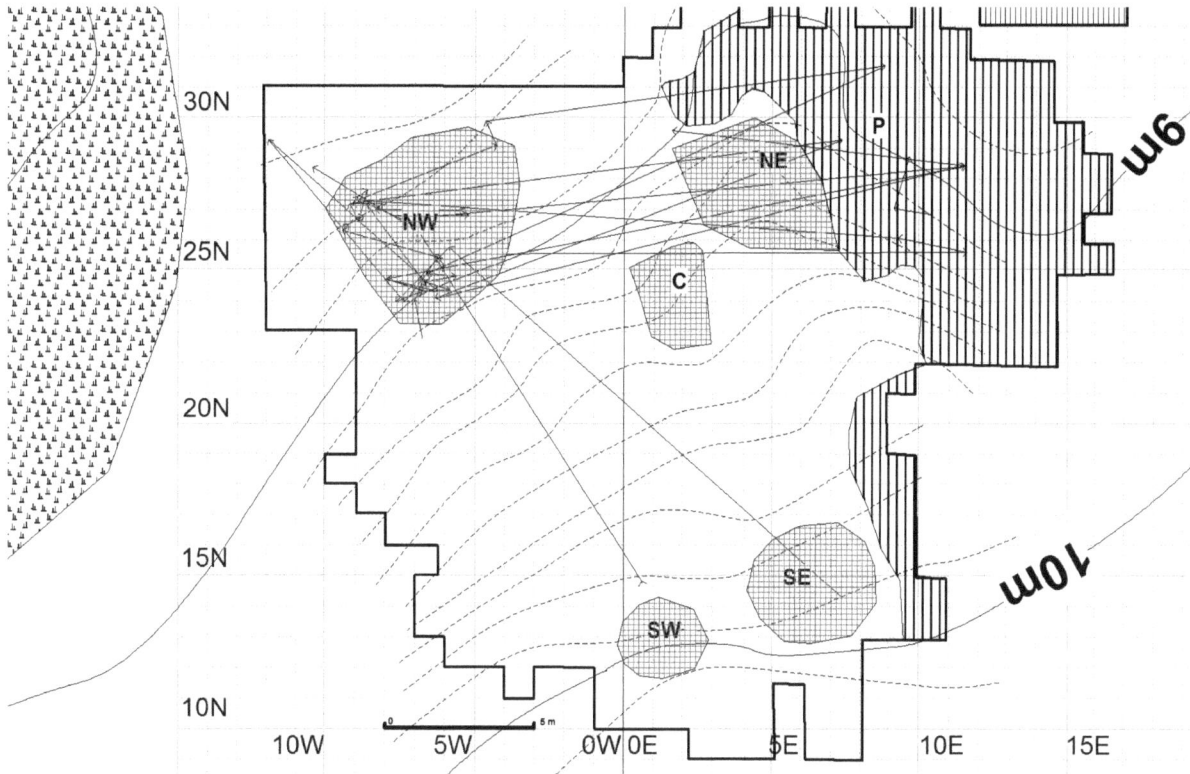

Figure 59 - Distribution of artefacts belonging to Refit 120-123.

Spatial layout: All elements were collected in the P-sector.

Refit 120-123

Refit state: A conjoinment of 38 blanks. Not all artefacts were refitted, but the raw material and the general form of the artefacts, especially the fact that nearly all blades and bladelets have a similar cortical edge, suggests that they belong to a single reduction sequence, which ended with two very small (< 3m long) exhausted opposed platform cores.

Original nodule: Flint type 5. Beneath the cortex there is a slightly darker coloured zone around 0.5 cm thick. Dim. min. 9 cm.

Reduction sequence: Blades and bladelets have been obtained apparently by the reduction of a rather flat nodule. Initially rather elongated blades but no traces of cresting are present.

Core/abandonment: 5.8 x 4.1 x 3 cm. One negative removal shows traces of hinging.

Productivity: It can be estimated that the sequence generated between 10 and 20 laminar products in addition to a range of trimming flakes. One of the long blades was selected for the production of proximal truncated blades (fig. 45, 2). One of the cores was terminated by a steeply plunging blade.

Spatial layout: The constituent elements were found scattered across the whole surface of the site (fig. 59).

Refit 124

Refit state: A conjoinment of 26 blanks. This refit comprises a near complete nodule, where only some decortication flakes seem to be absent (fig. 60)..

Original nodule: Flint type 8. Dim. min. 10 x 11 x 4 cm.

Reduction sequence: Debitage on this nodule started with the installation of a striking platform This procedure generated several large, mostly cortical flakes. Another striking platform was created on the opposite side of the nodule. Flakes are often plunging. Finally mainly large cortical flakes were produced, leaving two cores: an irregular core and a opposed platform core.

Figure 60 - Refit 124.

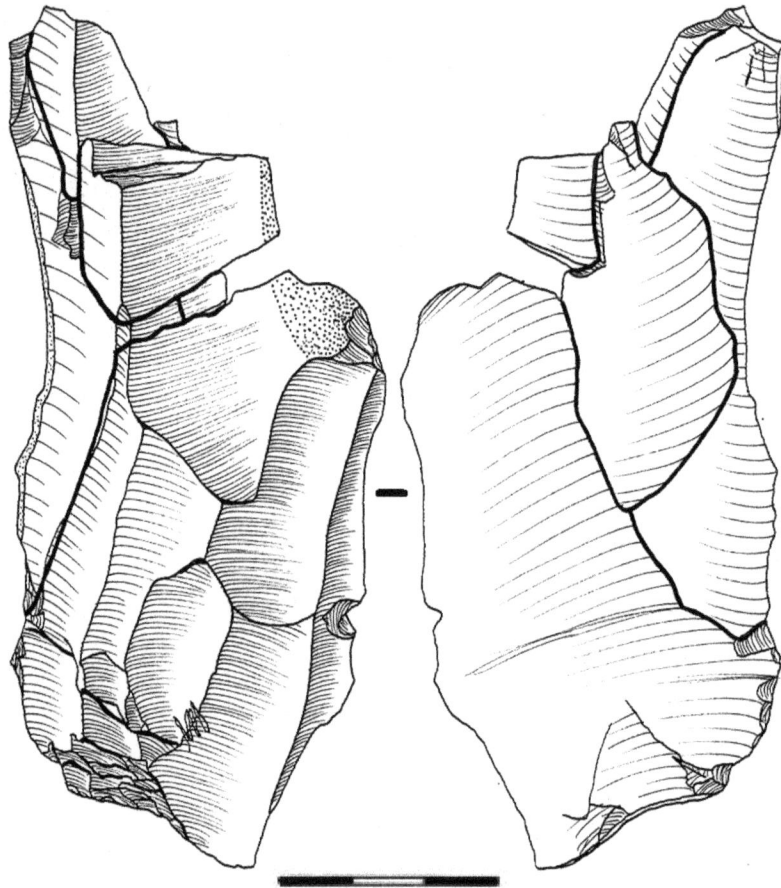

Figure 61 - Refit 127.

Core/ abandonment: A very irregular core and a exhausted opposed platform core.

Productivity: No good blades were obtained. Two small bladelets were produced, of which one was crested. Most products consist of cortical flakes and very irregular flakes, probably caused by the presence of frost cracks in the nodule. A small flake was used for the production of a Zonhoven point.

Quality of knapping: Poor production quality.

Spatial layout: Except for a bladelet from the P-sector all other elements come from the dense scatter in the NW-sector

Refit 125-128 (fig 47-48)

Refit state: This refit comprises a total of 45 flakes, blades and a core.

Original nodule: Flint type 8. (min length 12 cm). The nodule is of poor quality as the original, large nodule exploited for this sequence contained some flaws due to frost cracks.

Reduction sequence: The nodules were thinned by a few lateral preparations. Flaking proceeded from an opposed platform core. It immediately produced some cortical blades. The flaking direction was alternating. After some few removals the platform was renewed by a transversal tablet (fig. 47-48).

Productivity: Most removal are thick and not straight, apparently because of flaws in the nodule. Several flakes are

hinged.

Quality of knapping: No good blades were obtained.

Spatial layout: fig. 60.

Refit 131

Refit State: 6 flakes.

Original nodule : Flint type 10. Dim. unknown (min. dim. 7 cm).

Reduction sequence: A series of thick bidirectional flakes from an opposed platform core.

Core/abandonment: unknown.

Productivity: flakes.

Quality of knapping: Poor flake production.

Spatial layout: All elements were collected in the P-sector.

Refit 142

Refit State: 32 flakes and fragments conjoining with the core provide an accurate picture of the various reduction stages.

Original nodule: Flint type 10. Dim. 5 x 5 x 6 cm. The core was made on an ovoid frosted nodule the surface of which is intensely aeolised.

Figure 62 - Refit 142.

Reduction sequence: The first removals from this nodule arc cortical flakes, removing the top of the nodule and creating a single platform core (fig. 62). Successive cortical flakes were detached from around the nodule. Small lateral rectifications were produced, creating an unilateral crest, apparently to initiate the removal of a sequence of blades from the single platform. A hard hammer was used.

Core/abandonment: The core was exhausted, although further platform rejuvenation could have created a 'short' reduction face.

Productivity: Poor in terms of laminar output. Mainly large flakes. No good blades were obtained. The core was abandoned after the obtainment of some thick flakes, some of which are irregular and elongated.

Quality of knapping: mediocre.

Spatial layout: Most elements, core included, originate in the dense scatter from the NW-sector. Some flakes were scattered in and near the P-sector (fig. 61).

Refit 144

Refit state: 15 flakes and fragments conjoining with the core (fig. 64).

Original nodule: Flint type 8. Dim. 9 x 8 x 4 cm. A polyhedral frosted nodule was split along a frost crack. The other part is represented by refit 145.

Reduction sequence: In a first part the cracked, frosted surface served as the platform for a single platform core, which produced short bladelets on one side of the core and a plunging blade on the other. In the second part of the nodule the other frosted surface also used as a platform for the production of some short cortical bladelets and a single thick blade. Another natural frosted surface on the opposite side was used for the production of some short bladelets and a large plunging flake, which caused the abandonment of the core.

Core/abandonment: Opposed platform core.

Productivity: Poor in terms of laminar output.

Quality of knapping: mediocre.

Spatial layout: All elements, core included, were collected in the SE-sector.

Refit 145

Refit state: A conjoinment of 15 blanks with a single platform core (fig. 41, 5).

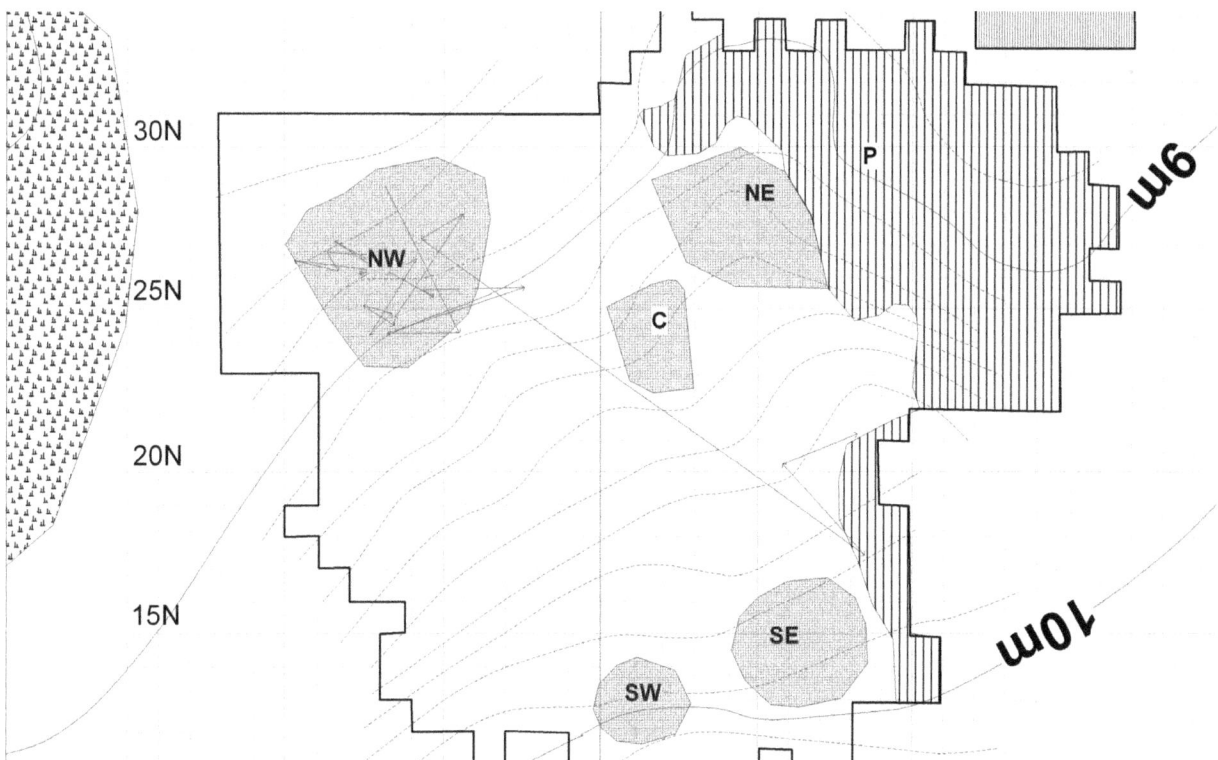

Figure 63 - Distribution of artefacts belonging to Refit 142.

Figure 64 - Refit 144 (upper part) and 145 (lower part).

Original nodule: A frosted angular nodule fragment of flint type 8 was used, which probably should refit with refit 144. Dim. unknown (min. length 7 cm).

Reduction sequence: Detachment of a cortical flake on top of the fragment created a first platform, which was used to produce some elongated flakes and a unilateral crested blade. Frost cracks resulted in flake fragmentation. The platform was rejuvenated by a transversal tablet and some more laminar flakes were obtained from the second platform.

Core/abandonment: Single platform core, entirely exhausted.

Productivity: cortical laminar flakes.

Refit 147 & 151

Refit State: A conjoinment of 10 flakes, illustrating part of a reduction sequence.

Original nodule: Frosted flint type 8.

Reduction sequence: Decortication flakes with numerous frost cracks from a single platform core.

Productivity: cortical flakes and flake fragments.

Quality of knapping: Very poor results.

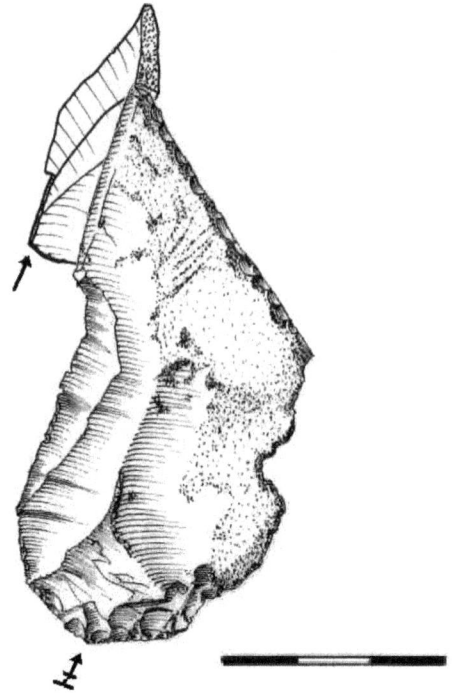

Figure 65 - Refit 156.

Spatial layout: All elements derive from the P-sector.

Refit 150

Refit state: A conjoinment of 5 elongated flakes, illustrating part of a reduction sequence.

Original nodule: Flint type 6. Dim. unknown (min. length 4 cm).

Reduction sequence: a sequence of unilateral flakes.

Productivity: cortical flakes.

Quality of knapping: good.

Spatial layout: All elements were collected in the NW-sector.

Refit 169

Refit state: A conjoinment of 9 blanks (fig. 66).

Original nodule: Frosted flint type 4. Dim. unknown (min. length 10 cm).

Reduction sequence: This set represents a sequence of bidirectional laminar reduction. There are clear traces of platform renewal.

Core/abandonment: Opposed platform core.

Productivity: A long plunging blade from a platform that was much higher than the present. Blades were obtained from the other platform but fractured because of a frost crack.

Quality of knapping: Unsuccessful laminar production because of the frosted flint.

Spatial layout: All elements derive from the P-sector.

Figure 66 - Refit 169.

Refit 170

Refit State: A conjoinment of 5 blanks (fig.40,3).

Original nodule: Flint type 4. Dim. unknown (min. length 6 cm).

Reduction Sequence: This set represents a sequence of bidirectional laminar reduction. Two tablets illustrate platform renewal. Reduction came to an end because of some hinged blade removals.

Core/abandonment: Opposed platform core.

Productivity: probably a series of good blades, which could not, however, be identified.

Quality of knapping: Successful laminar production. Several butts were intentionally prepared (facetted).

Spatial layout: All elements were collected in the NW-sector.

Refit 180

Refit State: A conjoinment of 12 blanks

Original nodule: Flint type 8. Dim. unknown (min. length 5 cm).

Reduction sequence: This set represents a sequence of a mainly bidirectional laminar reduction. There are clear traces of platform renewal.

Productivity: A series of good blades.

Quality of knapping: Successful laminar production. Most butts are flat but attest to platform abrasion.

Spatial layout: All elements were collected in the SE-sector, except for one from the P-sector.

Refit 182

Refit State: A conjoinment of 8 blanks

Original nodule: Flint type 8. Dim. unknown (min. length 9 cm).

Reduction sequence: This set represents a sequence of unidirectional laminar reduction. The first blade is unilaterally crested.

Productivity: A series of blades, all of them somewhat plunging.

Quality of knapping: Successful laminar production. Most butts are flat but attest to platform abrasion.

Spatial layout: All elements were collected in the SE-sector.

Refit 185

Refit state: A conjoinment of 6 blanks.

Original nodule: Flint type 8. Dim. unknown (min. length 7 cm).

Reduction sequence: This set represents a sequence of unidirectional laminar reduction. Two blade are unilaterally crested.

Productivity: A series of irregular blades.

Quality of knapping: Successful laminar production.

Spatial layout: All elements were found close to each other in the SE-sector.

Refit 199

Refit State: A conjoinment of 7 blanks (fig. 46,3).

Original nodule: Black flint type 1. Dim. unknown (min. length 10 cm).

Reduction sequence: This set represents a sequence of unidirectional laminar reduction following the rounding of the core.

Productivity: A series of cortical blades. The outer blade is unilaterally crested.

Quality of knapping: Successful laminar production.

Spatial layout: All elements, except one from the P-sector, were collected in the SE-sector.

Refit 208

Refit State: A conjoinment of 5 blanks

Original nodule: Flint type 8. Dim. unknown (min. length 9 cm).

Reduction sequence: This set represents a sequence of chunks and several flakes which were obtained using a frost fracture surface as a platform.

Productivity: A long series of nice blades.

Quality of knapping: Unsuccessful production.

Spatial layout: All elements were collected in the SE-sector.

Refit 209-210

Refit State: A conjoinment of 9 blanks

Original nodule: Flint type 8. Dim. unknown (min. length 7 cm).

Reduction sequence: This set represents a sequence of mainly cortical core preparation flakes. One of the flakes has been transformed into a multiple dihedral burin (fig. 51, 9).

Productivity: A series of flakes.

Quality of knapping: poor.

Spatial layout: All elements were collected in the SE-sector.

Refit 211

Refit State: A conjoinment of 7 blanks.

Original nodule: Flint type 8. Dim. unknown (min. length 7 cm).

Reduction sequence: This set comprises a large cortical flake, which was reworked into a burin before the removal of flakes. The flakes were detached from the proximal dorsal end of the cortical flake (fig. 51, 13). From the proximal right edge of the flake a transversal flake was detached, creating a surface from which burin spalls could be detached. On the right edge at least three burin spalls were detached, of which the second was mainly dorsal. On the left edge at least two plunging burin spalls were detached, creating a double dihedral edge burin.

Productivity: Flakes and burin spalls.

Quality of knapping: Poor.

Spatial layout: All elements were collected in the SE-sector and its environs.

Refit 212

Refit State: A conjoinment of 11 blanks

Original nodule: Flint type 8. Dim. unknown (min. length 7 cm).

Reduction sequence: This set represents a sequence a bidirectional flakes and bladelets. At least one platform renewal flake was produced (fig. 41, 5). Most bladelets originated from one of the platforms, while the other platform only produced flakes.

Core/abandonment: Opposed platform core, with preservation of the cortex at the back.

Productivity: A series of bladelets.

Quality of knapping: Successful laminar production.

Spatial layout: All elements were collected in the NW-sector.

Refit 220

Refit state: A conjoinment of 10 blanks.

Original nodule: Flint type 8. Dim. unknown (min. length 3 cm).

Reduction sequence: This set represents a decortication sequence of unidirectional flakes.

Productivity: A series of cortical flakes.

Quality of knapping: Poor

Spatial layout: All elements were collected in the SE-sector.

Refit 222

Refit state: A conjoinment of 5 blanks

Original nodule: Flint type 8. Dim. unknown (min. length 6 cm).

Reduction sequence: This set represents a sequence of frosted chunks and flakes.

Productivity: frosted chunks.

Quality of knapping: Unsuccessful flaking attempt.

Spatial layout: All elements were collected near to one other in the NW-sector.

Refit 222b

Refit state: A conjoinment of 5 blanks.

Original nodule: Flint type 8. Dim. unknown (min. length 6 cm).

Reduction sequence: This set represents a initial sequence of bidirectional flakes from an opposed platform core.

Productivity: A series of broken elongated flakes and a cortical blade with a unilateral crest.

Quality of knapping: Unsuccessful laminar production because of the presence of frost cracks.

Spatial layout: All elements were collected in the NW-sector.

Refit 227

Refit state: A conjoinment of 6 blanks.

Figure 67 - Refit 227.

Original nodule: Flint type 8. Dim. unknown (min. length 6 cm).

Reduction sequence: This set represents a sequence of uni-directional flakes (fig. 66). There is at least one platform renewal.

Core/abandonment: Single platform pyramidal core.

Productivity: A series of plunging blades.

Quality of knapping: Successful laminar production.

Spatial layout: All elements were collected in the P-sector.

Refit 235

Refit state: A conjoinment of 7 blanks

Original nodule: Flint type 8. Dim. unknown (min. length 7 cm).

Reduction sequence: This set represents a sequence in the production of flakes and blades from an opposed platform core made from a frost shattered nodule. There is platform renewal by the production of a core tablet.

Core/abandonment: Opposed platform core.

Productivity: A series of cortical bladelets, flakes an a single thick blade.

Quality of knapping: Poor.

Spatial layout: All elements were collected in the SE-sector.

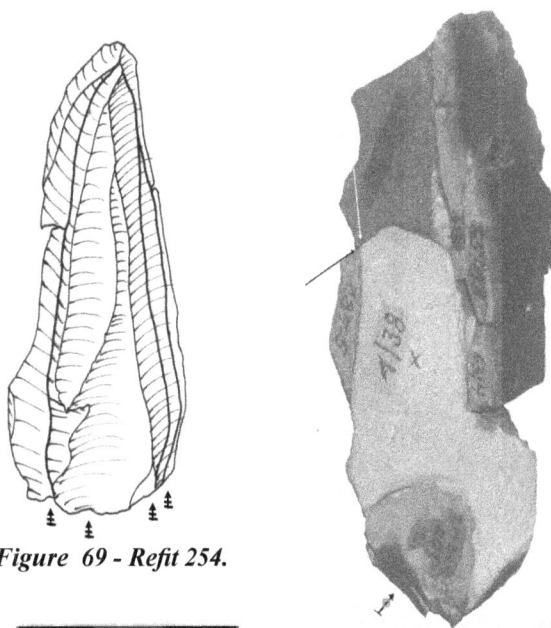

Refit 236

Refit state: A conjoinment of 7 blanks (fig. 68).

Original nodule: Flint type 8. Dim. unknown (min. length 7 cm).

Reduction sequence: This set represents a sequence in the production of (mostly broken) blades from an opposed platform core made from a frost affected nodule where a cortex hole limited the blade length. There is platform renewal by the production of a core tablet.

Productivity: A series of cortical and other blades.

Quality of knapping: Poor.

Spatial layout: All elements were collected in the SE-sector.

Refit 240 bis

Refit state: A conjoinment of 9 blanks.

Reduction sequence: This set represents a sequence of uni-directional flakes.

Productivity: A good series of blades.

Quality of knapping: Poor laminar production.

Spatial layout: All elements were collected in the NW, SE and P-sector.

Figure 68 - Refit 236.

Refit 255

Refit state: A conjoinment of 5 blanks

Original nodule: Flint type 1. Dim. unknown (min. length 8 cm).

Reduction sequence: This set represents a sequence of two unidirectional cortical flakes of which one is worked as a burin on a break with three successive burin spalls.

Productivity: A series of cortical flakes.

Quality of knapping: Good thin elongated cortical flakes.

Spatial layout: All elements were collected in the SE-sector.

Figure 69 - Refit 254.

Figure 70 - Refit 256.

73

Refit 256

Refit state: A conjoinment of 9 blanks.

Original nodule: Flint type 1. Dim. unknown (min. length 8 cm).

Reduction sequence: This set represents a sequence of bi-directional cortical blades. The largest blade obtained a series of the notches on the left edge. One of the notches was used to produce a edge dihedral burin (fig. 71).

Productivity: A good series of cortical blades and a burin.

Quality of knapping: Successful laminar production.

Spatial layout: All elements were collected in the NW-sector.

Refit 258

Refit state: A conjoinment of 8 blanks from a decortication sequence.

Original nodule: Flint type 1. Dim. unknown (min. length 7 cm).

Reduction sequence: This set represents a sequence of mainly unidirectional cortical flakes. One of the flakes was used for the creation of a double burin on a break. The platform from which the flakes were detached was formed by a transversal detachment of a critical blade. The blade was transformed into a retouched bitruncated blade (fig. 55, 11).

Productivity: A series of flakes.

Quality of knapping: Decortication sequence.

Spatial layout: All elements, except a flake from the NE-sector, were collected in the NW-sector.

Refit 259

Refit state: A conjoinment of 7 blanks.

Original nodule: Flint type 1. Dim. unknown (min. length 3 cm).

Reduction sequence: This set represents a sequence of multidirectional cortical flakes. On one of the cortical flakes a dihedral burin was created and the burin spall was used to make a burin on a break (fig. 71).

Productivity: Two cortical flakes.

Quality of knapping: Decortication sequence.

Spatial layout: All elements were collected in the NW-sector.

Refit 260bis

Refit state: A conjoinment of 5 blanks.

Original nodule: Flint type 1. Dim. unknown (min. length 5 cm).

Reduction sequence: This set represents a sequence of bi-directional laminar reduction. The last blade is plunging.

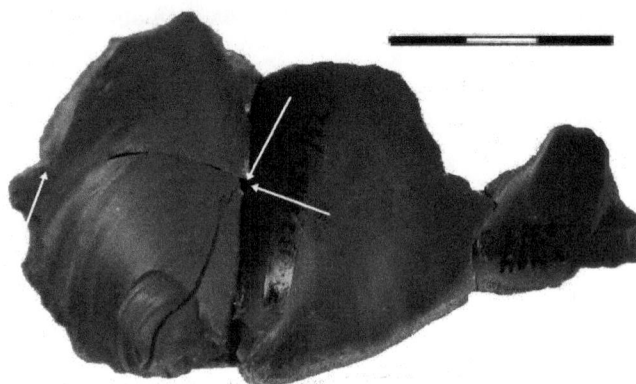

Figure 71 - Refit 259 dihedral burin and burin on a break.

Productivity: A series of blades.

Quality of knapping: Successful laminar production.

Spatial layout: All elements were collected in the SE-sector.

Refit 260q

Refit state: A conjoinment of 5 blanks.

Original nodule: Flint type 1. Dim. unknown (min. length 9 cm).

Reduction sequence: This set represents a sequence of two unidirectional laminar flakes.

Productivity: flakes.

Quality of knapping: Successful production.

Spatial layout: All elements were collected in the NW-sector.

Refit 261

Refit state: A conjoinment of 8 blanks.

Original nodule: Flint type 1. Dim. unknown (min. length 9 cm).

Reduction sequence: This set represents a sequence of a burin refreshing on a unilateral irregular blade. Initially, on the proximal end, a central dihedral burins was created of which four burins spalls were conjoined. At least one earlier burin spall was not found. At the distal end of the blade a ventral notch was created and a lateral steep retouch applied. This was used to section the blade and allowed the creation of a lateral dihedral burin (fig. 50, 12).

Productivity: A retouched blade used to form a multiple burin with several burin spalls.

Quality of knapping: Successful laminar production.

Spatial layout: All elements were collected in the SE and P-sector.

Refit 262

Refit state: A conjoinment of 8 blanks.

Original nodule: Flint type 1. Dim. unknown (min. length 9 cm).

Reduction sequence: This set represents a sequence of burins obtained on a large flake. The distal end of the flake has been retouched into an end scraper (or convex truncation). Some of the retouch chips were found. (fig. 50, 8, 9). On the scraper head a plunging burin blow, taking away an important part of the flake, created a edge burin on a convex truncation. Another burin blow on the burin facet created an edge burin on a break (fig. 50, 8). The rest of the flake after the plunging was used for manufacturing another double burin: burin on a break and dihedral burin (fig. 50, 10). Some spalls were refitted.

Productivity: A flake.

Quality of knapping: Successful laminar production. Several butts were intentionally prepared (facetted).

Spatial layout: All elements were collected in the NW-sector.

Refit 263

Refit state: A conjoinment of 7 blanks.

Original nodule: Flint type 1. Dim. unknown (min. length 7 cm).

Reduction sequence: This set represents a sequence of plunging blades together with two unilateral crested blades. The sequence is probably the result of a successive sharpening of a burin and the blades are thus probably burin spalls.

Productivity: Blades.

Quality of knapping: Laminar production.

Spatial layout: All elements were collected in the NW-sector.

Refit 265

Refit state: A conjoinment of 9 blanks.

Original nodule: Flint type 1. Dim. unknown (min. length 6 cm).

3. Reduction sequence: This set represents a sequence of burin sharpening comprising a dihedral burin and burin spalls.

Productivity: A dihedral burin.

Spatial layout: All elements were situated in the NW-sector.

Refit 266

Refit state: A conjoinment of 5 blanks.

Original nodule: Flint type 1. Dim. unknown (min. length 9 cm).

Reduction sequence: This set represents a sequence of blades from an opposed platform core. One of the blades has a distal straight truncation and an irregular retouch on the proximal left edge.

Productivity: Blades from opposed platform core.

Quality of knapping: Successful laminar production.

Spatial layout: All elements were collected in the NW-sector.

Refit 268

Refit state: A conjoinment of 11 blanks

Original nodule: Flint type 1. Dim. unknown (min. length 9 cm).

Reduction sequence: This set represents a sequence of two blades and a flake that have been broken. The first blade is the distal fragment of a unilateral crest. The left fragment of the flake has been reworked into a Zonhoven point (left middle centre on fig. 70).

Productivity: Several preparation flakes an unilateral crest. A fragment of a flake was changed into a Zonhoven point (left on figure 72).

Quality of knapping: Frosted chunk.

Spatial layout: All elements were collected in the NW-sector with the exception of the Zonhoven point, which was recovered in the NE-sector.

Refit 269

Refit State: A conjoinment of 6 blanks.

Original nodule: Flint type 1. Dim. unknown (min. length 8 cm).

Reduction sequence: This set represents a sequence of several blade fragments. (fig. 51, 7).

Productivity: At the distal end of the blade a dihedral edge burin was produced with at least four burin blows of which three spalls were found. The proximal end of the blade was splittered and a burin blow was applied.

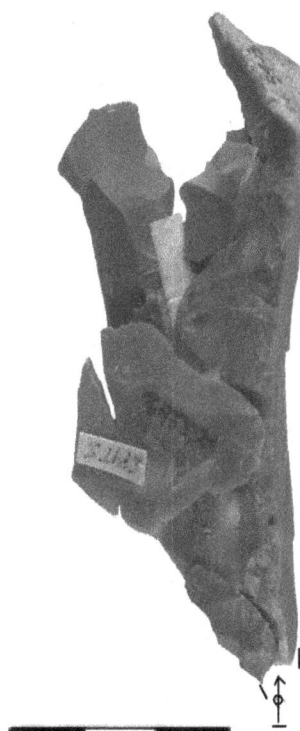

Figure 72 - Refit 268 including a Zonhoven point (central left).

75

The burin spall was plunging taking away most of the blade. Finally the plunging facet was used to produce a dihedral burin.

Quality of knapping: Nice blade.

Spatial layout: All elements were situated in the NW-sector.

Refit 271-275

Refit state: A conjoinment of 41 blanks.

Original nodule: Flint type 1. Dim. unknown (min. length 9 cm).

Reduction sequence: This set represents several fractured blades and blades from an opposed platform core. Initially some cortical blades were removed, mainly from one of the platforms. Nearly all flakes and blades are broken. There is no clear indication of frosting, which could be responsible for the fractured state of the reduction.

Productivity: blades, flakes and chips. One of the fragmented blades served for the manufacture of a burin on a concave truncation.

Quality of knapping: good.

Spatial layout: All elements were collected in the NW, SE and the SW-sectors. The artefact recorded from the SW-sector is a chunk.

Refit 274 quarter

Refit state: A conjoinment of 8 blanks.

Original nodule: Flint type 1. Dim. unknown (min. length 5 cm).

Reduction sequence: This set represents a sequence of uni-lateral and mainly cortical flakes.

Productivity: One of the flakes has been transformed into a dihedral burin from which three spalls were refitted.

Quality of knapping: Good. The burin spalls are affected by heat.

Spatial layout: All elements were collected in the NW-sector.

Refit 280

Refit state: A conjoinment of 4 blanks.

Original nodule: Flint type 8. Dim. unknown (min. length 4 cm).

Reduction sequence: This set represents a small cobble from which a tablet was obtained and some elongated flake. A unilateral crest was also created.

Core/abandonment: Single platform core.

Productivity: restricted.

Quality of knapping: good

Spatial layout: All elements were collected in the NW and SE-sectors.

Refit 281

Refit state: A conjoinment of 9 blanks.

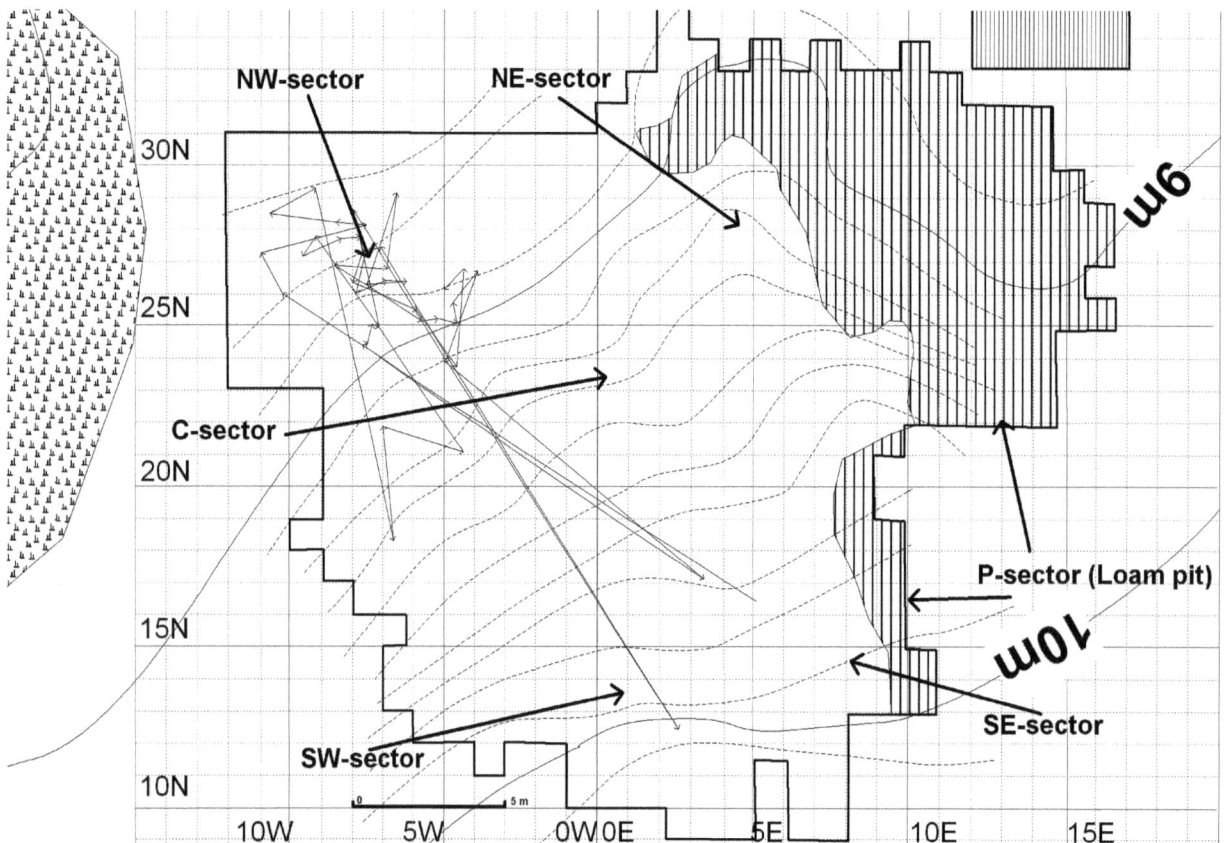

Figure 73 - Distribution of artefacts belonging to Refit 271-275.

Original nodule: Flint type 4. Dim. unknown (min. length 5 cm).

Reduction sequence: This set comprises a frost fractured nodule with a series of blades. The frosted surface served initially as a platform for the production of some flakes and blades which were not found. In opposed direction another platform was created from which several good cortical and other good blades have been obtained. No crest preparation is present.

Core/abandonment: A crossed platform core on a frost fragment.

Productivity: A series of blades.

Quality of knapping: Good quality with apparently no detrimental impact from the frosted condition of the nodule.

Spatial layout: All elements were collected in the NW, SE and NE-sectors.

Refit 282

Refit state: A conjoinment of 8 blanks.

Original nodule: Flint type 4. Dim. unknown (min. length 6 cm).

Reduction sequence: This set comprises a frost fractured nodule, possibly part of the same nodule as that of Refit 281. A platform was by flaking a series of chips from the frosted surface, transversal to the length axis of the nodule. Perpendicular to that surface a unidirectional crest was created, taking away part of the nodule cortex. From the platform several cortical and other blades were produced until the final plunging blade resulted in core abandonment.

Core/abandonment: Opposed platform core.

Productivity: blades.

Quality of knapping: Successful but limited laminar production.

Spatial layout: All elements were collected in the SE and P-sectors.

Refit 314

Refit state: A conjoinment of 5 blanks.

Original nodule: Flint type 8. Dim. unknown (min. length 7 cm).

Reduction sequence: This set represents a sequence of blades together with two unilateral crested blades from a core with platform renewal. A plunging blade, the distal end of which has a lateral preparation, was removed in order to obtain a convex core table.

Core/abandonment: Opposed platform core.

Productivity: three good blades.

Quality of knapping: Successful laminar production.

Spatial layout: All elements were collected in the P-sector.

Refit 324

Refit state: A conjoinment of 5 blanks.

Original nodule: Flint type 8. Dim. unknown (min. length 6 cm).

Reduction sequence: This set represents a sequence in which a frosted chunk was used for the creation of an opposed platform core, but broke along a frost crack.

Core/abandonment: chunk.

Productivity: frosted chunks.

Quality of knapping: Unsuccessful because of frosted condition of the core.

Spatial layout: All elements were collected in the SE-sector.

Refit 329

Refit state: A conjoinment of 5 blanks.

Original nodule: Flint type 3. Dim. unknown (min. length 6 cm).

Reduction sequence: This set represents a decortication sequence with lateral preparation producing only cortical flakes, some of which are broken because of the presence of frost cracks.

Productivity: cortical flakes.

Quality of knapping: poor.

Spatial layout: All elements were collected in the NW-sector.

Refit 331

Refit state: A conjoinment of 5 blanks.

Original nodule: Flint type 3. Dim. unknown (min. length 5 cm).

Reduction sequence: This set represents a sequence of decortication with the production of a Siret burin.

Productivity: cortical flakes.

Quality of knapping: poor.

Spatial layout: All elements were collected in the NE and P-sectors.

Refit 343

Refit state: A conjoinment of 10 blanks.

Original nodule: Flint type 9. Dim. unknown (min. length 5 cm).

Reduction sequence: This set represents a sequence of decortication (fig. 40, 9) during which the platform was renewed at least three times on one side and at least twice on the other side.
Core/abandonment: opposed platform core.

Productivity: mainly cortical flakes and some elongated

flakes.

Quality of knapping: poor quality.

Spatial layout: All elements were collected in the central SE-sector.

Refit 351

Refit State: A conjoinment of 10 blanks.

Original nodule: Flint type 9. Dim. unknown (min. length 5 cm).

Reduction sequence: This set represents a sequence of thick blades and elongated flakes obtained from a single platform core, together with a unilateral crested flake. The last blade was plunging, taking away an important part of the core.

Core/abandonment: single platform core.

Productivity: some elongated flakes and thick irregular blades.

Quality of knapping: poor quality.

Spatial layout: All elements were collected in the NW-sector.

Refit 356

Refit state: A conjoinment of 4 blanks.

Original nodule: Flint type 3. Dim. unknown (min. length 11 cm).

Reduction sequence: This set represents a sequence of frosted chunks which show clear traces of human retouch either to use it as a scraper or more probably as a single platform core.

Core/abandonment: all frosted chunks.

Productivity: some flakes and chunks.

Quality of knapping: very poor quality not producing usable blanks

Spatial layout: All elements were collected in the SE-sector.

Refit 361

Refit state: A conjoinment of about 50 blanks, refitted before the individual elements had been registered. It was decided not to break down the refit.

Original nodule: Flint type 8. Dim. unknown (min. length 13 cm).

Reduction sequence: This set represents a sequence of nodule decortication and preparation of an opposed platform core, On one side of the core a single platform was created, which was not, however, reworked. In contrast, on the other side several successive platforms were exploited (fig. 50).

Productivity: A significant number of nice, long blades were produced but several blades are fractured.

Quality of knapping: good but not excellent.

Spatial layout: The elements were collected mainly in the P-sector, but also in the SE sector.

Refit 396 bis

Refit state: A conjoinment of 5 blanks.

Original nodule: Flint type 8. Frosted nodule.

Reduction sequence: The nodule was splitted up into several frosted chunks of which one was used as an irregular core for the production of some small flakes and chips.

All elements were collected in the P-sector.

Refit 399

Refit state: A conjoinment of 7 blanks.

Original nodule: Flint type 3. Dim. unknown (min. length 10 cm).

Reduction sequence: This set represents a sequence of bi-directional nodule decortication and preparation of an opposed platform core, producing cortical flakes and blades.

Productivity: poor productivity of usable blades.

Quality of knapping: good.

Spatial layout: All elements were collected in the P-sector.

Refit 403

Refit state: A conjoinment of 20 blanks.

Original nodule: Flint type 8. Dim. unknown (min. length 5 cm).

Reduction sequence: This set represents a sequence of intensely frosted chunks, which show clear traces of human retouch. This is suggestive of multiple attempts to initiate a debitage session that were ultimately entirely unsuccessful.

7. Spatial layout: All elements were collected in the SE-sector.

Refit 404

Refit state: A conjoinment of 5 blanks.

Original nodule: Flint type 8. Dim. unknown (min. length 5 cm). A patinated flake which is frosted and has been used for producing flakes.

Reduction sequence: This set represents a sequence of intensively frosted blank.

Productivity: cortical flakes and a blade.

Quality of knapping: very poor quality not producing usable blanks.

Spatial layout: All elements were collected in the P-sector.

Refit 405

Refit State: A conjoinment of 7 blanks.

Original nodule: Flint type 8. Dim. unknown (min. length 5 cm).

Reduction sequence: This set represents a sequence comprising a frosted older patinated artefact which was split into two single platform cores for the production of some cortical flakes, which show clear traces of human retouch.

Core/abandonment: two single platform cores.

Productivity: cortical flakes and chips.

Quality of knapping: very poor quality not producing usable blanks.

Spatial layout: All elements were collected in the P-sector.

Refit 406

Refit state: A conjoinment of 21 blanks.

Original nodule: Flint type 8. Dim. unknown (min. length 10 cm).

Reduction sequence: This set represents a sequence of unidirectional nodule decortication with unilateral crest and preparation of a single platform core.

Core/abandonment: single platform core.

Productivity: good quality blades from a single platform core. Linear butt and reduced bulb.

Quality of knapping: good quality.

Spatial layout: All elements, except one from the NW-sector, were collected in the P-sector.

Refit 410

Refit state: A conjoinment of 8 blanks.

Original nodule: Flint type 8. Dim. unknown (min. length 5 cm).

Reduction sequence: This set represents a sequence of flakes and blades with unilateral crest preparation from an opposed platform core.
Productivity: rather thick blades. Quality of knapping: poor quality.

Spatial layout: All elements were collected in the SE-sector.

Refit 481

Refit state: A conjoinment of 6 blanks

Original nodule: Flint type 1. Dim. unknown (min. length 7 cm).

Reduction sequence: This set represents a sequence of three fragmented thin cortical blades from a single platform core. One of the proximal blade fragments has been burnt, after fracture.

Productivity: nice cortical blades.

Quality of knapping: good quality.

Spatial layout: All elements were collected in the P-sector.

Conclusions regarding the knappers competence

The raw material used was generally of poor to very poor quality. Often only a few flakes or blades were obtained from the frost cracked nodules. The purpose was mainly to produce blades/bladelets. The knappers did not invest significant preparation activity in the initial decortication but nevertheless were able to obtain some good blades, albeit often cortical. Continuous blade production often started by the preparation of a crest, mostly unifacial. During blade production frost cracks present in the nodule often resulted in the fracturing of blades. Only occasionally could a long sequence of nice blades be obtained. Often plunging ended blank production.

An important exception to this general knapping procedure is provided by black flint type 1. No specialised flaking techniques were used and the bulbs are generally well pronounced. Butts are most often flat. Most blanks seem to have been obtained by a hard hammer technique. In this flint type large blades and flakes were obtained.

6 - FUNCTIONAL ANALYSIS OF THE LITHICS

by Veerle Rots

A selection of tools from the site of Zonhoven-Molenheide was submitted to a functional study as part of a Master's thesis (Rots 1996) at *Katholieke Universiteit Leuven* (KU-Leuven), Belgium. Analysis was performed using an Olympus binocular reflected-light microscope (magnification 50-500x) and was based on an experimental reference collection available at KULeuven. Unfortunately, a large part of the assemblage proved to be heavily affected by post-depositional alterations, which greatly restricted reliable use-wear identification. Nevertheless, the few results that could be obtained are summarised below.

Tools were selected for analysis from each of the different typological categories. From a total assemblage of 98 generic tools available at the time (excluding microliths and microblades), a sample of 61 tools was analysed. This sample was randomly selected per typological category. Wear traces proved to be identifiable on only 16 tools. The majority (11), consisting primarily of burins, proved to have been used for bone working, while 5 tools (mainly scrapers) showed evidence of dry hide working.

The archaeological assemblage available at the time also included of 41 microblades, 12 of which were examined microscopically. All proved to be too heavily affected by post-depositional alterations to allow for any functional determination.

For the microliths breakage patterns were also studied. The definitions and classifications used were based on those proposed by the Ho Ho Committee (1979), which are also reproduced in Fischer (1984). Step-terminating bending fractures, preferably in association with spin-off fractures and microscopic linear impact traces (MLIT's) were considered to be diagnostic of a use of the microlith as a projectile. From a total of 61 microliths, 31 pieces remained complete while 30 showed various types of fracture. At least 8 microliths from the fracture group could be classified as projectiles based on sufficiently reliable evidence. Typologically, these were mainly points with oblique truncation (7 from 21 examined), but there was also a curved backed point (1 from 3 examined). While other types of bending fractures were noted (i.e., snap, feather, hinge), these were not considered to be sufficiently diagnostic to identify the point as having been part of a projectile arrangement.

In conclusion, the functional analysis of the lithics from Zonhoven-Molenheide only provided minimal and very partial results. The analysis provided reliable evidence for hunting activities and the processing of animal material at the site, but no insights could be obtained into the processing of other materials, either for subsistence purposes or maintenance activities. Therefore a definitive conclusion regarding the site's function is not possible.

References

Fischer A., Hansen P. V. , *et al.* 1984. Macro and Micro Wear Traces on Lithic Projectile Points. *Journal of Danish Archaeology* 3: 19-46.

Ho Ho Committee 1979. The Ho Ho Classification and Nomenclature Committee Report. Lithic use-wear analysis. B. Hayden. Academic Press.

Rots V. 1996. *Gebruikssporenonderzoek op de lithische artefacten van het site te Zonhoven-Molenheide 2. Aanvullend experimenteel onderzoek* Unpublished Ma thesis, Katholieke Universiteit Leuven.

7 - DISCUSSION

7.1 - Stratigraphic position of Ahrensburgian sites

In many other sandy areas of Western Europe, one encounters stratigraphic problems similar to those encountered at Zonhoven-Molenheide. This is principally due to the very restricted area over which natural sedimentation contemporaneous or posterior to the Ahrensburgian took place. With the exception of a few cave deposits in upland areas, West European sites are mainly open air sites situated in, or on, Weichselian aeolian deposits. Such conditions repeatedly result in palimpsest situations for Holocene human remains.

Unmixed Late Palaeolithic, Mesolithic and Neolithic occupation remains could only be preserved in those specific areas where Younger Dryas dune building or peat formation has occurred and/or where volcanic tuffs have been deposited. Such dune building can be identified in the field when Usselo soil, of Allerød age, is preserved below Holocene soil horizons (Vanmontfort *et al.* 2010). As the stratigraphy of the aeolian deposits is mostly obliterated by the Holocene soil, there is often no possibility of identifying Younger Dryas deposits. Indeed, only rarely can a specific deposit of Younger Dryas age be identified and differentiated from other earlier deposits.

When no discrete Late Pleistocene or Holocene deposits are present, any occupation remains belonging to these periods will have been subjected to bioturbation in the Holocene soil, which results in a vertical distribution of the artefacts in soil horizons (Barton 1987; Vermeersch, Bubel 1997). For that reason, the stratigraphic position of artefacts is often of no use for the construction of site chronologies. Such problems are well illustrated by figure 73 (Vermeersch 2011) and certainly apply to the site of Zonhoven-Molenheide.

7.2 - Homogeneity of the sector assemblages

At the site of Zonhoven-Molenheide several artefact concentrations are present. When analysing the distribution of artefacts across the site, it became apparent that not all artefact types shared the same distribution pattern (fig. 21-35, 38-39). This could be interpreted as the result of a palimpsest situation. In such a hypothesis it would be probable that artefacts from different occupation events, characterised by different assemblage compositions, would have formed in different concentrations, as is indeed the case with the various concentrations that were identified in different sectors at the site.

Raw material and typological considerations point to at least the presence of an Ahrensburgian and a Mesolithic occupation at the site of Zonhoven-Molenheide.

One should, however, be very careful when interpreting intra-site horizontal patterns of archaeological materials. Insights will be increased if it can be shown that the remains belong to a single occupation. But, how can one be sure? A very small occupation scatter may be an indication that the site has been occupied for only a very restricted period. Such an artefact scatter could be the result of a specialised activity. In such situations only a limited amount of the total material variability may be present, thus limiting the scope of inter-site typological and technological comparison.

At the site of Zonhoven-Molenheide it is not possible to make a definitive, clean split between artefacts belonging to an older Ahrensburgian and a younger Mesolithic occupation. Occupation remains from other visits to the site might also be present and there is always the possibility of superposition of multiple prehistoric occupations at the same location. One always faces the possibilities of a superposition of several prehistoric occupations at the same spot. So far we have no methods that might allow us to split up such multi-phase collections.

At other sandy sites, where multiple phases of occupation are suspected, certain methods have been employed to split up an assemblage into older and younger components. For example, at Brecht-Moordenaarsven 2 (Vermeersch Lauwers, Gendel 1992), emphasis was placed on the depth of the artefacts within the soil horizons, proceeding from the hypothesis that younger artefacts should be found at a higher level because they had less time to migrate down into the soil. While the results produced by this approach seemed useful, one can never be entirely certain about the real value of the results produced by such a division.

At the site of Zonhoven-Molenheide a division based on the vertical distribution of the artefacts could not be made because the precise position of artefacts was not recorded in three dimensions. There are thus no stratigraphic data which could validate or invalidate a division of the material between Ahrensburgian and Mesolithic and consequently, it is difficult to make judgements about the homogeneity of the excavated material. Nevertheless certain observations at Zonhoven-Molenheide could suggest the presence of a mixture of different occupation remains.

First, there is the presence of charcoal dating to at least two separate periods of time. The 14C date UtC-3720 (10760 ± 70 BP) is statistically different from that of UtC-3195 (7060 ± 70 BP). These dating results suggest the presence of at least two occupation periods: a late Palaeolithic and a Mesolithic. This suggestion is confirmed by the presence of a collection of artefacts in Wommersom quartzite, a raw material that was apparently not often used during the Late Palaeolithic (Gendel 1984).

Second, there is the presence of some geometric microliths, such as symmetric trapezes, segments and triangles and probably also most of the backed bladelets (fig. 52), which might more easily be assigned to a Mesolithic assemblage. However, the impact of Mesolithic artefacts on the whole assemblage, in much the same way as that of the Wommersom quartzite, seems to be restricted.

It is possible to draw a distribution plan of the items that on typological grounds could be attributed to a Mesolithic occupation (fig. 74). This plan suggests that "Mesolithic" items are mainly found in the P-sector and are clearly overrepresented in the C-sector. They are underrepresented in the NW, SW and NE-sectors.

Since many burins and blades are made from the black flint type 1 (fig. 39), it is tempting to conclude that this material has been introduced by the Late Palaeolithic visitors to the site. The distribution of this material is clearly overrepresented in the NW and SE sectors and also indicates links between the NW and P-sectors. Such a distribution would suggest that the NW, SE and P-sectors belong together and are the result of a single occupation event. However some

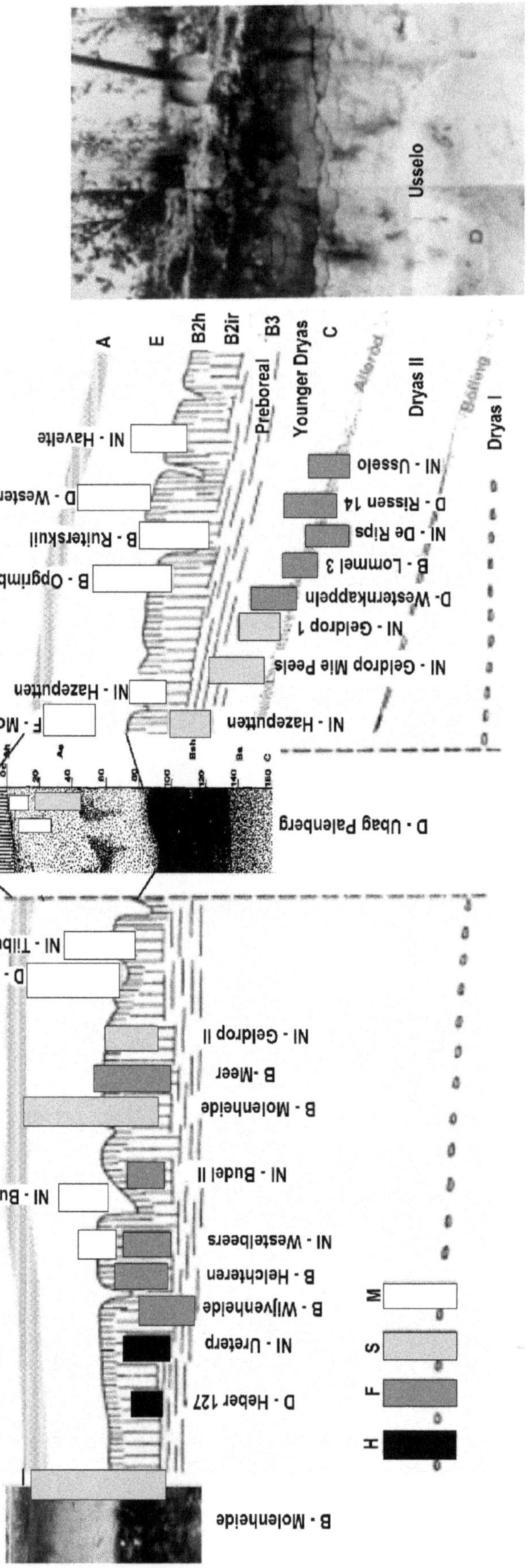

Figure 74 - Pedostratigraphic position of Late Glacial and Holocene sites (H: Hamburgian; F: Federmesser; A: Ahrensburgian; M: Mesolithic) in sandy deposits in western Europe (adapted from Vermeersch 1977, 2011 and Bosinski 1995).

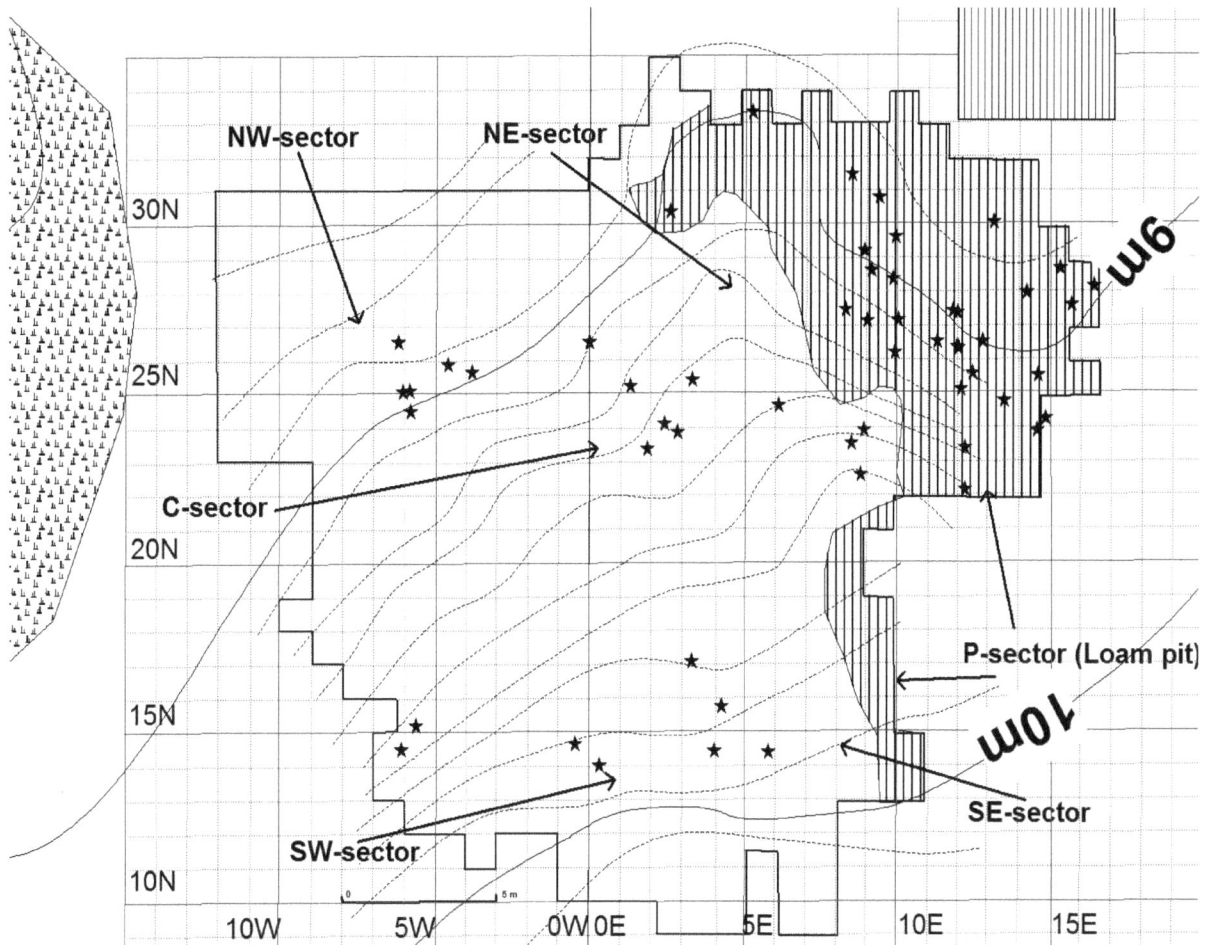

Figure 75 - Distribution of possible Mesolithic artefacts (i.e. Wommersom quartzite and some microliths).

rather "Mesolithic"-looking artefacts have also been found in these three sectors.

Accepting all the restrictions that pertain to such an approach, a tentative tool list belonging to a single, putative Late Palaeolithic occupation can be assembled (table 4). As such, this list is not fundamentally different from the list of all retouched blanks (fig. 3) and is thus unlikely to improve our understanding of what is, or is not, diagnostic of the Ahrensburgian.

7.3 - The Ahrensburgian of the Low Countries

The Ahrensburgian is part of the "Stielspitzen Gruppe" defined by W. Taute (1968) for Middle Europe. The 'classic' Ahrensburgian sites from the end of Younger Dryas or even the earliest Preboreal, such as Stellmoor (Rust 1943), are characterized by a predominance of tanged points. Taute (1968: 220) also included several sites with scarcely any tanged points in the Ahrensburgian, essentially because

Table 4 - Possible Late Palaeolithic tool list.

General type	%
scraper	4
burin	7
backed point	4
backed bladelet	9
Zonhoven-Ahrensburg point	43
microlith	6
truncation	10
miscellaneous	11

of the presence of large blades, including "Riesenklingen". He suggested that such sites might represent a late phase of the Ahrensburgian, dating from the last part of the Younger Dryas or the first part of the Preboreal (Taute, 1968: 21). Gob (1988; 1991) has included such sites in his Epi-Ahrensburgian.

In the Benelux and in western Germany most sites (exceptions being Vessem-Rouwven (Arts, Deeben 1981), Budel IV and Neer III (Bohmers 1956), have assemblages where Zonhoven points are much more numerous than Ahrensburgian points. Ahrensburgian points may even be rare or absent (tab. 5). However, such assemblages are still characterised by a good quality blade technology, the presence of numerous burins and end scrapers.

According to T.B. Ballin and A. Saville (2003) many Ahrensburgian tanged points were manufactured using a microburin technique. In the Benelux and in western Germany the Ahrensburgian is characterised by the near absence of the microburin technique. The individuality of the Ahrensburgian of Zonhoven-Molenheide has been stressed by De Bie (1999), who observes that percussion marks on Ahrensburgian blanks can be distinguished from both Federmesser and early Mesolithic assemblages.

We would like to define the local Ahrensburgian as a lithic industry that may be differentiated both from the Federmesser and from the Early Mesolithic by the production of large blades, characterised by specific percussion marks, the absence of the microburin technique, the (near) absence of (pointed) backed blades, the presence of numerous Zonhoven points and/or Ahrensburgian points,

Table 5 - Tool composition from Ahrensburgian sites, except Schulen 1, which is early Mesolithic (adapted from different sources): Schulen 1: Lauwers, Vermeersch 1982; Eersel: Deeben, Dijkstra, Van Gisbergen 2000; Remouchamps: Dewez 1987, Geldrop Mie Peels: Deeben, Dijkstra, van Gisbergen 2012; Vessem Rouwven: Arts, Deeben 1981, Oudehaske and Gramsbergen: Johansen and Stapert 1997-98; Geldrop 1, 3-1 and 3-2: Deeben 1994-1997; Zonhoven-Kapelberg: Huyge 1985, 1986.

	Zonhoven Molenheide	Zonhoven Kapelberg	Vessem Rouwven	Remouchamps	Geldrop 1	Geldrop 3-1	Geldrop 3-2 Oost	Geldrop Mie Peels	Eersel Panberg	Oudehaske	Gramsbergen 1	Schulen 1
All tools	384	123	772	94	121	256	300	50	74	61	164	187
All Microburins	0	0	10	0	0	0	2	0	1	5	0	84
Endscraper	26	27	178	11	22	68	54	4	20	12	43	49
Borer	1	0	10	06	1	1	4	0	0	0	1	4
Burin	28	21	109	9	15	32	31	6	7	7	29	22
Backed bladelet	52	0	52	8	4	13	1	0	0	4	2	89
Zonhoven point	165	49	49	17	45	51	127	319	16	31	76	39
Ahrensburg point	8	0	320	5	16	19	1	0	0	1	1	0
Trapeze (not Mesolithic)	8	1	3	4	4	7	7	1	1	3	8	1
Triangle	12	0	2	3	0	5	1	0	0	0	0	5
miscelanneous	48	24	50	11	14	54	73	30	28	3	10	28

the scarcity of geometric microliths and the presence of numerous burins and end-scrapers. Points sometimes represent up to 50 % of the retouched artefacts.

The absence or scarcity of Ahrensburgian points at some sites (Johansen, Stapert 1997-98) could be the result of the specific role of the site in the settlement system, or because reindeer were not exploited at that location or were not eaten during the season of the occupation (Deeben, Schreurs 2012). Alternatively, it might indicate a local chronological characteristic that is also expressed in the scarcity of the microburin technique.

7.4 - The Ahrensburgian chronology

According to Terberger (2004) the Ahrensburgian should have developed in the late Allerød in northern Germany, where such early assemblages do not exhibit Federmesser elements. The Ahrensburgian hunter-gatherers might have been more common in northern areas at times when there were more favourable living conditions than those prevailing during the Younger Dryas. Indeed, at the site of Alt Duvenstedt LA 121 in Schleswig Holstein, an artefact layer with unquestionable Ahrensburgian finds was detected in a soil horizon that was assigned to the Allerød (Kaiser, Clausen 2005). Among the artefacts were three scrapers, two burins and five small tanged points. There are also two AMS-dates from charcoal from fireplaces: 10810 ± 80 (AAR-2245-1) and 10770 ± 80 BP (AAR-2245-2). These place the occupation at the end of the Allerød and situate the site around 10.65 cal BC, which is much older than most other Ahrensburgian sites. But here also, as with many other sites, the question arises as to whether the charcoal is the product of human Ahrensburgian activity or from a non-human fire in during earlier Allerød times. The presence of the charcoal in a hearth seems to indicate the first possibility, but can the second possibility be excluded?

An analysis of the available ^{14}C dates (fig. 75) from Ahrensburgian assemblages in Benelux and Germany indicates that many are younger than the Younger Dryas (Lanting, Mook 1977; Vermeersch 2012) and demonstrates that in most cases problems of reliability exist. These problems are mainly related to the stratigraphical position of the assemblages (cf. 7.2) and to the related question of whether or to what extent the artefacts and the dated material are coeval. Moreover, nearly all sites attributed to the Ahrensburgian, and most sites attributed to the Epi-Ahrensburgian, present problems regarding the homogeneity of their assemblages.

Furthermore, the quantity of ^{14}C dates currently available is very restricted and can certainly not be accepted as a fully representative sample of Ahrensburgian chronology.

These considerations also apply to the site of Zonhoven-Molenheide.

7.5 - Internal site organisation

The Ahrensburgian site Geldrop 3-2 East is characterised by a rectangular, find-rich central region around a fire pit, which is surrounded by a loose scattering of artefacts. There can be no doubt regarding the reality of this pattern. In the artefact distribution the internal concentration emerges as a rectangular structure, even though no evidence for a constructed wall was found. Although the construction of such a boundary need not require a lot of work, its presence does at least suggest that the occupants were planning to remain in the same place for some time. The rich assemblage from this site produced many indications of household activities, including arrow hafting, which appears to have been one of the main activities. The rich assemblage, the demarcated central area and the long distance refits all suggest that the site has been used for a long period of time (Wenzel 2009).

Analysis of the distribution of artefacts and features at Geldrop Mie Peels suggested that a row of ochre patches could be interpreted as a windbreak. Such a windbreak, made of hides that had been processed using red ochre, would have divided the settlement into a sleeping area and a refuse area (Deeben, Schreurs 2011).

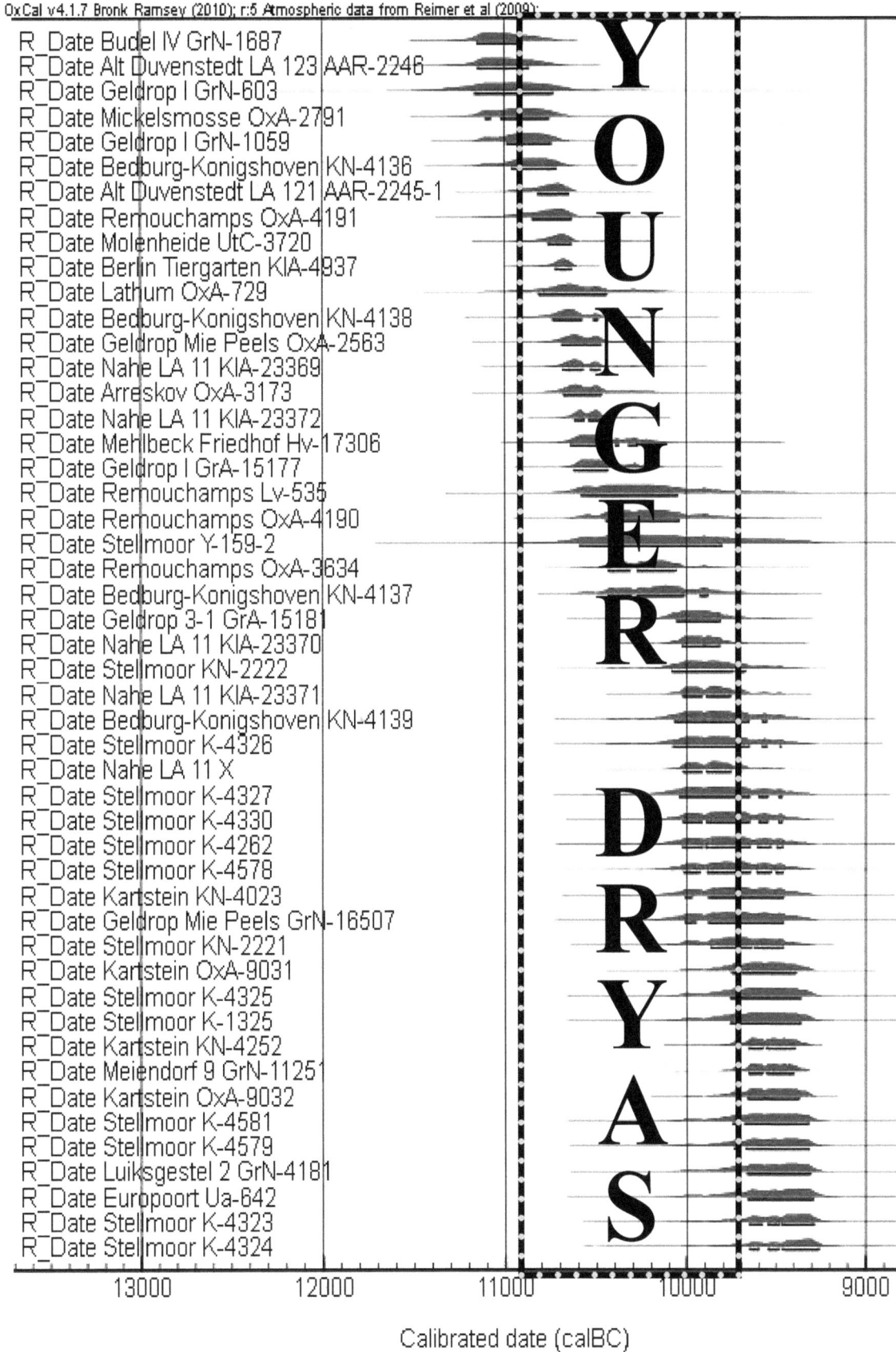

Figure 76 - List of calibrated ¹⁴C and AMS dates from sites in Western Europe attributed to the Ahrensburgian or the Epi-Ahrensburgian. Dates older than 11100 BP and younger than 9900 BP have been omitted from the list (listed after Vermeersch 2012). The time period of the Younger Drays (10850-9650 calBC) is indicated.

Table 6 - List of available ^{14}C-dates in uncalibrated BP from sites attributed to the Ahrensburgian in Germany and Benelux, restricted to dates older than 9500 and younger than 11000 BP (Vermeersch 2012).

Site, Lab reference	BP	±
Bedburg-Konigshoven, KN-3883B	9540	120
Meiendorf 9, GrN-11253	9550	40
Bedburg-Konigshoven, KN-3883A	9660	120
Stellmoor, OxA-2875	9680	90
Europoort, Ua-644	9690	125
Bedburg-Konigshoven, KN-4135	9740	100
Geldrop 3-2 East, GrA-15182	9770	60
Eersel-Panberg, GrA-15175	9810	70
Kartstein, KN-4254	9900	45
Europoort, Ua-642	9945	115
Luiksgestel 2, GrN-4181	9970	105
Kartstein, OxA-9032	9995	65
Meiendorf 9, GrN-11251	10000	40
Kartstein, KN-4252	10000	50
Stellmoor, K-4325	10010	100
Kartstein, OxA-9031	10020	75
Stellmoor, KN-2221	10080	80
Bedburg-Konigshoven, KN-4139	10140	100
Nahe LA 11, KIA-23371	10142	49
Stellmoor, KN-2222	10160	90
Nahe LA 11, KIA-23370	10172	45
Geldrop 3-1, GrA-15181	10190	60
Bedburg-Konigshoven, KN-4137	10290	100
Remouchamps, OxA-3634	10320	80
Remouchamps, OxA-4190	10330	110
Geldrop I, GrA-15177	10500	70
Mehlbeck, Friedhof, Hv-17306	10515	95
Nahe LA 11, KIA-23372	10544	49
Arreskov, OxA-3173	10600	100
Nahe LA 11, KIA-23369	10610	80
Geldrop, Mie Peels, OxA-2563	10610	100
Bedburg-Konigshoven, KN-4138	10670	100
Lathum, OxA-729	10670	160
Berlin Tiergarten, KIA-4937	10730	40
Molenheide, UtC-3720	10760	70
Alt Duvenstedt LA 121, AAR-2245-2	10770	60
Remouchamps, OxA-4191	10800	110
Alt Duvenstedt LA 121, AAR-2245-1	10810	80
Bedburg-Konigshoven, KN-4136	10920	100
Mickelsmosse, OxA-2791	10980	110

At Zonhoven Molenheide the site organisation is similar to that of the Federmesser sites at Rekem (De Bie, Caspar, 2000) and Meer (Van Noten, 1978). Artefact distribution is characterized by a wide dispersal (> 100 m²), with diversified lithic concentrations of 30 to 50 m².

At Zonhoven-Molenheide several sectors with important artefact concentrations have been identified over an area of about 500 m². No surface was free from remains and, moreover, it should be noted that a large area of the site has been destroyed by later digging activities. While it is clear that all sectors are interconnected by refits, no data were collected that might suggest how the different sectors were related. It remains unclear if standing structures were present during the Ahrensburgian occupation. The presence of a single post at the eastern edge of the SW-sector is difficult to interpret as its contemporaneity with the lithic artefacts could not be established. The large cobbles, whether concentrated or scattered (fig. 17), which lie mainly below the artefact rich horizon, seem not to have played a role in the structuring of space within the settlement.

Tool distribution over the excavated area indicates that the NW-sector was the most important area, where many activities were performed. Flaking took place in the NW, SE and P-sectors, as suggested by the distribution of cores, cortical artefacts and rejuvenations. Burins appear to have been most often discarded in the NW-sector, whereas they were most often sharpened in the SE-sector; the P-sector also playing a role. Scrapers are especially common in the SW-sector, which may suggest that some of them are of Mesolithic age.

Zonhoven points are characteristic for the NW and P-sectors, which could be an indication that retooling took place in those sectors. It is also worth noting that the charcoal, which is probably coeval with the late Palaeolithic, was collected from the NW-sector and may thus be related to the need for fire for retooling. However, burnt artefacts are more frequent in the SE-sector and also the C-sector, where no charcoal was collected. Truncations are apparently related to the (Zonhoven) points.

Backed pieces, found mainly outside the "Ahrensburgian sectors", could correspond to an area, the C-sector, where the impact of later Mesolithic visitors was greatest.

7.6 - Zonhoven-Molenheide in a regional context

The regional Ahrensburgian has recently been studied by J. Deeben (1994-1999), who looked at sites from the Southern Netherlands, and M.l Baales (1996), who has given an overview of the Ahrensburgian in the upland area. P.M. Vermeersch (2008, 2011) and M.-J. Weber, S.B. Grimm and M. Baales (2011) have studied the Ahrensburgian presence during the Younger Dryas.

In the region around Zonhoven-Molenheide, a number of sites have been attributed to the Ahrensburgian (De Bie & Vermeersch 1998). The most important ones are Vessem-Rouwven (Arts, Deeben 1981), Geldrop 1 and 2 (Deeben 1994), Geldrop 3-1 (Deeben 1995), Geldrop 3-2 Oost (Deeben 1996) and Geldrop Mie Peels (Deeben, Dijkstra, van Gisbergen 2002, Deeben and Schreurs 2012). Other sites (Arts & Deeben 1981) are present but are too poorly documented. There are also several caves: the Cave of the Coléoptère (Dewez 1987: 399-428), Fonds-de-Forét (Dewez 1987: 307-323; Gob 1988: 261-264), the Cave of the Préalle (Dewez 1987: 432-434) and Remouchamps (Dewez et al. 1974; Dewez 1987: 345-363).

The nearby site of Zonhoven-Kapeldreef (Huyge 1985, 1986) could also belong to the Ahrensburgian. Indeed, the assemblage (tab. 5) is very similar to that of Zonhoven-Molenheide.

Certain sites from the Netherlands are very rich in lithic equipment (Arts, Deeben 1981) but are sometimes poorly informed or even falsified such as the site of Vessem XII (Wouters 1996).

Generally speaking it can be deduced from the list of radiocarbon dates (fig. 75) that most of the sites are coeval with the Younger Dryas. The current best estimates place the Younger Dryas between 12,900 and 11,700 cal BP (Straus, Goebel 2011).

The absence at Molenheide of deposits attributable to the Younger Dryas might have caused the deflation of the human-produced charcoal. Some of the charcoal eventually migrated down into the lower soil horizons. A charcoal sample in such a stratigraphical position could be dated and gave an acceptable but not ideal result because of the problems discussed above. The charcoal sample, collected at 25.07N 05.69W at an elevation of 8.19m in the B3-horizon from the NW-sector has been dated by AMS to 10760 ± 40 (UtC-3720). Another charcoal sample from 28N 06W at an elevation of 8.39m in the B2ir-horizon, also from the NW-sector, gave a date of 7060 ± 70 (UtC-3195). There can be no doubt that the dates refer to two separate periods. The oldest date has the best fit with the technology and typology of the collected assemblages and will therefore be considered as the best indicator for the site's period of occupation.

The oldest Ahrensburgian sites in the area lack numerous Ahrensburgian points, but always have a high number of Zonhoven points (tab. 5). It is remarkable that in the Benelux Zonhoven points from the Ahrensburgian are not manufactured with the microburin technique, while this apparently is the case at more northern sites (Bali, Saville 2002). The microburin technique only first appears in the Benelux from the Early Mesolithic onwards (Vermeersch 1984), as exemplified by the site of Schulen 1 (tab 5) (Lauwers, Vermeersch 1982).

The Ahrensburgian assemblage of Zonhoven-Molenheide may be distinguished from both Federmesser and early Mesolithic assemblages. In the case of the latter, this ability to differentiate is surprising, given that microliths and retouched tool types generally seem to indicate continuity between the Ahrensburgian and the Early Mesolithic (Gob 1988).

It would seem that the oldest Ahrensburgian occupations are not typical for the northern European position of the sites. . In addition, it would seem that the so-called Epi-Ahrensburgian sites should not be considered to be clearly of younger age than sites that are "typical Ahrensburgian". The difference between them does not seem to relate to chronological or environmental differences. We prefer to consider it as the result of site specialisation or other personal preferences of the hunters.

According to P. Petitt and M. White (2012), finds of Ahrensburgian points in south eastern Britain, although rare, indicate that groups of this cultural affiliation were also operating there in this period. Further to the north and west, where evidence of a contemporary presence is available, such sites are lacking.

There is no doubt that the Young Dryas period in western Europe is still badly understood. As has already been noted, most sites from Western Europe that may be attributed to the Ahrensburgian are characterised by a chronology that is far from well established. The analysed dataset (Vermeersch 2012) comprises 184 ^{14}C and AMS dates for the period 10.1-11.0 ka BP that coincides with the Younger Dryas. These dates derive from sites in Germany, the Benelux and France, north of 49.5 N and west of 11° E, and from the U.K. east of 4° W. The dataset attributes only 31 dates to the Ahrensburgian, 4 to the Long Blade Complex (Belloisian), 52 to the Federmesser (Azilian, Creswellian), 45 to unspecified Late Palaeolithic and 6 to Magadalenian. How many of those dates are reliable?

7.7 - The landscape during the period of occupation

In the NW European lowlands, the Late Dryas was marked by renewed landscape instability and (extensive and rapid) aeolian activity as a result of climatic cooling, increased aridity and an opening up of the vegetation cover.

At the Allerød/Late Dryas transition and possibly even during the Late Dryas (Vanmontfort et al., 2010), one or more phases marked by forest fires took place at the investigated localities, as shown by the presence of macroscopic charcoal particles at the top of the Usselo horizon (Van der Hammen, Van Geel, 2008; Derese et al. 2011).

It is supposed that during the first part of the Younger Dryas, discontinuous permafrost was present in the Netherlands (Isarin, Bohncke 1999) and in northern Belgium, where frost wedges of Younger Dryas age have frequently been observed. The periglacial mounds (lithalsas) from the Hautes Fagnes in the Ardennes, which result from the activity of segregation ice, were formed during the Younger Dryas. They imply an annual average temperature of between -4 and -6°C (Pissart, 2003). The presence of cryofractured limestone fragments in the Ahrensburgian sediments of Remouchamps and La Préalle sites (Dewez et al. 1974) also suggest the presence of a very cold climate. Considering these climatic conditions and keeping in mind that human populations had deserted the Benelux during the LGM and returned to the area only during the Bølling, around 13200 cal BP (Vermeersch et al. 1987; Vermeersch, Maes 1996), one is tempted to conclude that the adverse environmental conditions that prevailed during the Younger Dryas were not particularly attractive for humans.

7.8 - Economic basis of the site

It is important to observe that Federmesser sites are considerably more numerous than Ahrensburgian sites. The Allerød interstadial conditions are characterised by a stable landscape, which apparently was very attractive to Federmesser groups. During the Younger Dryas there is a clear drop in of settlement when compared to the Allerød (Crombé, Verbruggen 1999). This could be explained by a thinning out of population or even a depopulation of the area. Further to the south J.-P. Fagnart and P. Coudret (1997) have observed that no Federmesser (Azilian) sites in the Somme basin in northern France could be attributed to the Younger Dryas. The occupation of that region thus appears to have declined sharply during the Younger Dryas (Fagnart 2009).

The Kempen landscape became less attractive during the Younger Dryas, resulting in a significant reduction of visits by human hunter-gatherers. Based on data from sites in the area of the river Somme, it would seem that red deer, roe deer and elk were replaced by reindeer at the onset of the Younger Dryas (Baales, Street 1999).

Seasonal patterns in the way Ahrensburgian groups used the landscape may be interpreted as essentially determined by seasonal movements of reindeer (Baales 1996). In this model, the main summer grazing areas of reindeer were in upland areas, such as the Ardennes and the uplands of central England, while during the autumn reindeer migrated to the southern part of the North Sea. Sites such as Remouchamps and Kartstein were occupied during the warmer season, when the reindeer migrated to the south, while lowland sites were occupied during the autumn and winter, when the reindeer moved north. Under this model the Zonhoven-Molenheide site, and many other sites in the Netherlands, may be viewed as places where Ahrensburgian groups halted for a short period during their transit from one region to the other. This would suggest that the occupation period coincided with either the spring or autumn. Unfortunately seasonal indicators at these sites are absent. No bones or other organic material were collected at Zonhoven-Molenheide.

The location of the site on the edge of the Kempen Plateau offered the possibility of exploiting two different landscapes: that of the dry Kempen Plateau and that of the sandy wetlands to the West. Immediately adjacent to the site a spring allows water to seep to the surface, which certainly would have been a welcome feature for a camp.

8 - CONCLUSIONS

The Younger Dryas was certainly a period of very cold conditions with a reinstallation of permafrost and significant deforestation. Reindeer made their reappearance in the area. The important Federmesser group, characterising the earlier Allerød period, seems to have disappeared at the onset of the Younger Dryas. However, it remains possible, if not clearly attested, that part of the Federmesser group continued its occupation. Another group, the Ahrensburgian, occupied and expanded in the lowlands and uplands. This occupation was probably on a seasonal basis, with winters spent in the lowlands, an area that also includes the present North Sea, while summers were spent in the Southern uplands. However, the Ahrensburgian has a rather sparse presence and was certainly not as extensive as the Federmesser groups. Regionally, the Ahrensburgian can be defined typologically as an assemblage with numerous Zonhoven points, a variable number of Ahrensburgian points and numerous end scrapers and burins. Irregular trapezes may be present. Technologically, the Ahrensburgian is differentiated from the Federmesser by the presence of long blades and by the absence of the microburin technique. At the end of the Younger Dryas and in the early Preboreal the whole of the present Benelux, northern France, the dry North Sea and Southern England was used by an Ahrensburgian group. We see no reason to differentiate between an Ahrensburgian and an Epi-Ahrensburigan group. The origin of the Ahrensburgian remains entirely unclear. The late Ahrensburgian groups gave birth to the early Mesolithic and a continuous occupation of the Benelux.

In order to deepen our understanding of the Ahrensburgian we need to look for new sites with faunal remains, interpretable stratigraphies and good dating opportunities. The possibility of finding such sites is, however, very restricted. In the sandy lowland deposits such sites are certainly rare. It would be advantageous if we could locate such sites in a peaty environment. A new cave site could also provide important new data.

9 - REFERENCES

Armour-Chelu M., Andrrews P. 1994. Some effects of bioturbationby earthworms (Oligocaeta) on archaeological sites. *Journal of Archaeological Science* 21: 433-443.

Arts N., Deeben J. 1981. *Prehistorische jagers en verzamelaars te Vessem: een model.* Stichting Brabants Heem.

Atkinson R.L.C. 1957. Worms and weathering. *Antiquity* 31: 219-33.

Baales M. 1996. *Umwelt und Jagdökonomie der Ahrensburger Rentierjäger im Mittelgebirge.* Verlag des Römisch Germanischen Zentralmuseums.

Balek C.L. 2002. Buried Artifacts in Stable Upland Sites and the role of Bioturbation: A Review. *Geoarchaeology: An International Journal* 17: 41-51.

Bali T.B., Saville A. 2003. An Ahrensburgian-type tanged point from Shieldaig, western Ross, Scotland, and its implications. *Oxford Journal of Archaeology* 22/2: 115-131.

Barton R.N.E. 1987. Vertical Distribution of Artefacts and Post-Depositional Factors Affecting Site Formation. In: P. Rowley-Conwy, M. Zvelebil and H.P. Blankholm (eds.), *Mesolithic Northwest Europe: Recent Trends.* University of Sheffield: 55-62.

Barton R.N.E. 1992. *Hengisthury Head, Dorset. Volume 2: The Late Upper Palaeolithic & Early Mesolithic sites.* Oxford University Committee for Archaeology Monograph 34.

Bohmers A. 1956. Statistics and graphs in the Study of Flint Assemblages. I. Introduction. II. A Preliminary Report on the Statistical Analysis of the Younger Palaeolithic of Northern Europe. *Palaeohistoria* 5: 1-26.

Bosinski G. 1995. Palaeolithic Sites in Rheinland. In: W. Schirmer (ed.), *Quaternary field trips in Central Europe.* München: 829-1000.

Bronk Ramsey C. 2009. Bayesian analysis of radiocarbon dates. *Radiocarbon,* 51: 337-360.

Cornwall I.W. 1958. *Soils for the Archaeologist.* Phoenix House.

Crombé P., Verbruggen C. 2002. The Lateglacial and early Postglacial occupation of northern Belgium: the evidence from Sandy Flanders. In: B.V. Eriksen, B. Bratlund (eds.) *Recent studies in the Final Palaeolithic of the European plain.* Jutland Archaeological Society: 165-180.

Crombé P. 1993. Tree-fall features on final Palaeolithic and Mesolithic sites situated on sandy soils : how to deal with it. *Helinium* 28: 50-66.

Cziesla E. 1988. Über das Kartieren von Artefakten-mengen in steinzeitlichen Grabungsflächen. *Bulletin de la Société préhistorique luxembourgeoise* 10: 5-53.

De Bakker H., Edelman-Vlam A.W. 1976. *De Nederlandse bodem in kleur.* Stichting voor Bodemkartering.

De Bie M., Vermeersch P.M. 1998. Pleistocene-Holocene Transition in Benelux. *Quaternary International* 49/50: 29-43.

De Bie M. 1999. Knapping techniques from the Late Palaeolithic to the Early Mesolithic in Flanders (Belgium): preliminary observations. In : A. Thévenin and P. Bintz (eds.), *L 'Europe des derniers chasseurs, Epipaléolithique et Mésolithique, Peuplement et paléoenvironnement de l'Epipaléolithique et du Mésolithique.* Editions du CTHS: 177-199.

De Bie M., Caspar J. 2000. *Rekem. A Federmesser Camp on the Meuse River Bank.* I.A.P., Leuven University Press.

Deeben J. 1994. De Laatpaleolithische en Mesolithische sites bij Geldrop (N.Br.), Deel 1. *Archeologie* 5: 3-57.

Deeben J. 1995. Human occupation of the southern Netherlands during the Younger Dryas (extended abstract). *Geologie en Mijnbouw* 74: 265-269.

Deeben J. 1995. De Laatpaleolithische en Mesolithische sites bij Geldrop (N.Br.), Deel 2. *Archeologie* 6: 3-52.

Deeben J. 1996. De Laatpaleolithische en Mesolithische sites bij Geldrop (N.Br.), Deel 3. *Archeologie* 7: 3-79.

Deeben J. 1997. De Laatpaleolithische en Mesolithische sites bij Geldrop (N.Br.), Deel 4. *Archeologie* 8: 33-68.

Deeben J. 1999. De Laatpaleolithische en Mesolithische sites bij Geldrop (N.Br.), Deel 5. *Archeologie* 9: 3-35.

Deeben J., Dijkstra P., Van Gisbergen P. 2000/2001. Nieuwe ^{14}C-dateringen van de Ahrensburg-cultuur in Zuid-Nederland. *Archeologie* 10: 5-19.

Deeben J., Schreurs J. 2012. The Pope, a miracle and an Ahrensburgian windbreak in the municipality of Waalre (province of Noord-Brabant), the Netherlands. In: M.J.L.Th. Niekus, R.N.E. Barton, M. Street, Th. Terberger (eds), *A mind set on flint. Studies in honour of Dick Stapert.* Groningen, Groningen Archaeological Studies 16: 295-318.

Derese C., Vandenberghe D., Van Gils M., Mees, F. Paulissen E., Van den Haute P. 2012. Final Palaeolithic of the Campine region (NE Belgium) in their environmental context: optical age constraints. *Quaternary International* 251: 7-21.

Dewez M. 1987. *Le Paléolithique Supérieur Récent dans les Grottes de Belgique.* Louvain-La-Neuve, Publications d'Histoire de l'Art et d'Archéologie de l'Université Catholique de Louvain 57.

Dewez M., Brabant H., Bouchud J., Callut M., Dambion F., Degerbol M., Ek C., Frère H., Gilot E. 1974. Nouvelles Recherches â la grotte de Remouchamps. *Bulletin de la Société Royale Belge d'Anthropologie et de Préhistoire* 85: 5-161.

Fagnart J.-P., Coudret P., 1997. Les industries á Federmesser dans le bassin de la Somme: chronologie et identité des groupes culturels. *Bulletin de la société préhistorique française* 84: 348-360.

Fagnart J.-P., 2009. Les industries â grandes lames et éléments mâchurés du paléolithique final du nord de la France: une spécialisation fonctionnelle des sites épi-ahrensbourgiens. In: Ph. Crombé, M. Van Strydonck, J. Sergeant, M. Boudin, M. Bats (eds.), *Chronology and Evolution Within The Mesolithic of North-West Europe: Proceedings of An International Meeting, Brussels, 30e June 2007*. Cambridge Scholars Publishing: 39-56.

Frederickx E., Gouwy S. 1996. *Toelichting bij de Quartairgeologische Kaart.* Vlaamse overheid, Dienst Natuurlijke Rijkdommen.

Gendel P.A. 1984. *Mesolithic Social Territories in Northwestern Europe.* BAR International Series 218.

Gob A. 1988. L'Ahrensbourgien de Fonds-de-Fôret et sa place dans le processus de mésolithisation dans le nord-ouest de l'Europe. In: M. Otte (ed.), *De la Loire â l'Oder. Les civilisationg du Paléolithique final dans le nord-ouest européen.* BAR International Series 444 : 259-285.

Gob A. 1991. The Early Postglacial Occupation of the Southern Part of the North Sea Basin. In: N. Barton, A.J. Roberts, D.A. Roe (eds.), *The Late Glacial in North-West Europe: Human Adaptation and Environmental Change at the End of the Pleistocene.* CBA Research Report 77: 227-233.

Gullentops F., Bogemans F., De Moor G., Paulissen E., Pissart A. 2001. Quaternary lithostratigraphic units (Belgium). *Geologica Belgica* 4: 153-164.

Huyge D. 1985. Een Vroeg-Epipaleolithisch wooncomplex te Zonhoven-Kapelberg. *Limburg* 64: 183-202.

Huyge D. 1986. Een Vroeg-Epipaleolithisch wooncomplex te Zonhoven-Kapelberg (Belgisch Limburg). *Notae Praehistoricae* 6: 29-32.

Inizan M.L., Reduron-Ballinger M., Roche H., Tixier J. 1999. *Technology and Terminology of Knapped Stone.* Nanterre.

Inizan M.L., Roche H., Tixier J. 1992. *Technology of Knapped Stone.* Meudon.

Isarin R.F.B., Bohncke S.J.P. 1999. Mean July temperatures during the Younger Dryas in Northwestern and Central Europe as inferred from climate indicator plant species. *Quaternary Research* 51: 158-173.

Johansen L., Stapert D. 1997-1998. Two 'Epi-Ahrensburgian' Sites in the Northern Netherlands: Oudehaske and Gramsbergen. *Palaeohistoria* 39/40: 1-87.

Kaiser K., Clausen I. 2005. Palaeopedology and stratigraphy of the lae Palaeolithic Alt Duvenstedt site, Schleswig-Holstein (Northwest Germany). *Archäologisches Korrespondezblatt* 35: 447-466.

Lanting J.H., Mook W.G. 1977. *The Pre- and Protohistory of the Netherlands in terms of Radiocarbon date*s. Groningen.

Lauwers R., Vermeersch P.M. 1982. Mésolithique ancien â Schulen. In: P.M. Verneersch (ed.) *Contributions to the Study of the Mesolithic of the Belgian Lowland.* Studia Praehistorica Belgica 1: 55-112.

Munaut A.V. 1967. Recherches paléo-écologiques en Basse et Moyenne Belgique. Leuven, *Acta Geographica Lovaniensia* 6.

Narr K.J. 1968. *Studien zur Älteren und Mittleren Steinzeit der Niederen Lande.* Rudof Habelt Verlag.

Paulissen E. 1973. *De morphologie en de kwartair-stratigrafie van de Maasvallei in Belgisch Limburg.* Koninklijke Academie voor Wetenschappen, Letteren en Schone Kunsten van België.

Paulissen E. 1983: Les nappes alluviales et les failles quaternaires du Plateau de Campine. In: F. Robaszynski, C. Depuis, *Guides Géologiques Régionaux: Belgique.* Masson: 167-170.

Peleman C., Vermeersch P.M. 2002. Zonhoven: Ahrensburg nederzetting. Limburg-Het Oude Land van Loon 81: 317320.

Peleman C., Vermeersch P.M., Luypaert L. 1994. Ahrensburg Nederzetting te Zonhoven-Molenheide 2, *Notae Praehistoricae* 14: 73-80.

Petitt P., White M. 2012. *The British Palaeolithic: Human Societies at the Edge of the Pleistocene World.* Routledge.

Pissart A. 2003. The remnants of Younger Dryas lithalsas on the Hautes Fagnes Plateau in Belgium and elsewhere in the world. *Geomorphology* 52: 5-38.

Rots V. 1996. *Gebruikssporenonderzoek op silexartefacten van de Ahrensburg-nederzetting te Zonhoven.* Unpublished M.A. thesis, K.U.Leuven.

Rozoy J.G. 1978. *Les derniers chasseurs. L'Epipaléolithique en France et en Belgique. Essai de synthèse.* Charleville, 3 vols.

Rust A. 1943. *Die Alt- und Mittelsteinzeitlichen Funde von Stellmoor.* Archäologisches Institut des Deutschen Reiches. Karl Wachholtz Verlag.

Scheys G., Dudal R., Bayens L. 1954. *Une interprétation de la morphologie de podzols humo-ferriques.* Transactions. Fifth International Congress Soil Science Léopoldville 4: 274-281.

Schwabedissen H. 1954. *Die Federmesser-Gruppen des nordwesteuropaïschen Flachlandes, zur Ausbreitung des Spät-Magdalénien.* Offa-Bücher.

Schwantes G. 1928. *Nordisches Paläolithikum und Mesolithikum – Festschrift.* Museum für Volkerkunde.

Stein J.K. 1983. Earthworm Activity: A Source of Potential Disturbance of Archaeological Sediments. *American Antiquity* 48: 277-289.

Straus L., Goebel T. 2011. Humans and Younger Dryas: Dead end, short detour, or open road to the Holocene? *Quaternary International* 242: 259-261.

Taute W. 1968. *Die Stielspitzen-Gruppen im Nördlichen Mitteleuropa. Ein Beitrag zur Kenntnis der späten Altsteinzeit.* Fundamenta A5, Böhlau Verlag.

Terberger T. 2004. The Younger Dryas - Preboreal transition in northern Germany - facts and concepts in discussion. In: T. Terberger, B.V. Eriksen (eds.), *Hunters in a changing world. Environment and Archaeology of the Pleistocene -*

Holocene Transition [ca. 11000 - 9000 B.C.] in Northern Central Europe. Verlag Marie Leidorf: 203-222.

Van der Hammen, T., Van Geel B. 2008. Charcoal in soils of the Allerød-Younger Dryas transition were the result of natural fires and not necessarily the effect of an extraterrestrial impact. *Netherlands Journal of Geosciences (Geologie en Mijnbouw)* 87: 359-361.

Van Noten F. 1978. *Les Chasseurs de Meer.* Dissertationes Archaeologicae Gandenses, 18, De Tempel.

Vanmontfort B., Van Gils M., Paulissen E., Bastiaens J., De Bie M., Meirsman E. 2010. Human settlement in the Late- and Early Post-Glacial environments of the Liereman Landscape (Campine, Belgium). *Journal of Archaeology in the Low Countries.* jalc/02/nr02/a02.

Vermeersch P.M. 1976. La position lithostratigraphique et chronostratigraphique des industries épipaléolithiques et mésolithiques en Basse Belgique. *Congrès préhistorique de France - XXe Session, Provence (1974)*: 616-621.

Vermeersch, P.M. 1977. Die stratigraphische Probleme der Postgiazialen Kulturen in Dünengebieten. *Quartär* 27/28: 103-109.

Vemeersch P.M., 1984. Du Paléolithique final au Mésolithique dans le nord de la Belgique. In: D. Cahen, P. Haesaerts (Edts.) *Peuples chasseurs de la Belgique Préhistorique dans leur cadre naturel.* Bruxelles: 181-193

Vermeersch P.M. 1991. TL Dating of the Magdalenian Sites at Orp, Belgium. *Notae Praehistoricae* 10, 27-29.

Vermeersch P.M. 1999. Postdepositional Processes on Epipalaeolithic and Mesolithic Sites in the Sandy Area of Western Europe. In : A. Thévenin (ed.) *L'Europe des derniers chasseurs.* Editions du CTHS: 159-166.

Vermeersch P.M. 2008. La transition Ahrensbourgien-Mésolithique ancien en Campine, Belgique en dans le Sud sableux des Pays Bas. In: J.-P. Fagnart (ed.), *Le début du Mésolithique en Europe du Nord-Ouest.* Mémoire XLV de la Société préhistorique française: 11-29.

Vermeersch P.M. 2011. The human occupation of the Benelux during the Younger Dryas. *Quaternary International* 242: 267-276.

Vermeersch P.M. 2012. Radiocarbon Palaeolithic Europe Database v14. http://ees.kuleuven.be/geography/projects/14c-palaeolithic/.

Vermeersch P.M., Bubel S. 1997. Postdepositional artefact scattering in a podzol: Processes and consequences for late Palaeolithic and Mesolithic sites, *Anthropologie* (Brno) 35/2-3: 119-130.
Vermeersch P.M., Creemers G. 1994. Early mesolithic sites at Zonhoven-Molenheide. *Notae Praehistoricae* 13: 63-69.

Vermeersch P.M., Lauwers R., Gendel P. 1992. The Late Mesolithic sites of Brecht-Moordenaarsven (Belgium). *Helinium* 32: 3-77.

Vermeersch P.M., Maes M. 1996. Chronostratigraphy of the Magdalenian at Orp. *Notae Praehistoricae* 16: 87-90.

Vermeersch P.M., Peleman C., Maes R. 1998. De Ahrensburg nederzetting te Zonhoven-Molenheide. *Limburg - Het Oude Land van Loon* 77: 51-52.

Vermeersch P.M., Peleman C., Rots V., Maes R. 1996. The Ahrenburgian site at Zonhoven-Molenheide. *Notae Praehistoricae* 16: 117-121.

Vermeersch P.M., Symens N., Vynckier P., Gijselings G., Lauwers R., 1987. Orp, site magdalénien de plein air (Commune d'Orp-Jauche). *Archaeologia Belgica* III: 7-56.

Washburn A.L. 1973. *Periglacial processes and environments.* Edward Arnold.

Weber M.-J., Grimm S.B., Baales M. 2011. Between warm and cold: Impact of the Younger Dryas on human behavior in Central Europe. *Quaternary International* 242: 277-301.

Wenzel S. 2009. *Beahusungen im Späten Jungpaläolithikum und Mesolithicum Nord-, Mittel- und Westeuropas.* Verlag des Römisch-Germanischen Zentralmuseums.

Wood W, Johnson D. 1978. A Survey of Disturbance Processes in Archaeological Site Formation. In: *Advances in Archaeological Method and Theory* 1: 315-351.

Wouters A.M. 1996. Een nieuwe intacte vindplaats van de Ahrensburgkultuur, tussen de "Aardhorst" en het "Rouwven": Vessem XII. *Apan/Extern* 5: 64-66.

www.ingramcontent.com/pod-product-compliance
Lightning Source LLC
Chambersburg PA
CBHW061010030426
42334CB00033B/3430